INSTITUT FRANÇAIS DU PÉTROLE
ÉCOLE NATIONALE SUPÉRIEURE DU PÉTROLE ET DES MOTEURS

Robert P. MONICARD
Coordinator for Field Production
Industrial Direction
Institut Français du Pétrole

PROPERTIES OF RESERVOIR ROCKS: CORE ANALYSIS

Translation from the French
by David BERLEY

1980

ÉDITIONS TECHNIP 27 RUE GINOUX 75737 PARIS CEDEX 15 technip

ISBN 2-7108-0387-9

Printed in France
by Imprimerie Louis-Jean, 05002 Gap

AMONG OUR BOOKS

Publication in English

- Enhanced Oil Recovery
  M. Latil
  with the assistance of C. Bardon, J. Burger, P. Sourieau

- The Reservoir Engineering Aspects of Fractured Formations.
  L.H. Reiss

- Multiphase Flow in Porous Media.
  C.M. Marle

- Seabed Reconnaissance and Offshore Soil Mechanics for the Installation of Petroleum Structures.
  P. Le Tirant

- Drilling Data Handbook.

PREFATORY NOTE

Specialists from the petroleum industry, the *Institut Français du Pétrole (IFP),* the *Ecole Nationale Supérieure du Pétrole et des Moteurs (ENSPM)* and universities teach at the Graduate Study Center for Drilling and Reservoir Engineering of *ENSPM.*

In conjunction with this teaching, they have written various books dealing with the different scientific and technical aspects of these petroleum operations.

The present book, which is part of the series of production courses written under the guidance of A. Houpeurt, is one of them.

contents

INTRODUCTION: Reservoir rock characteristics objective of core analysis and general presentation of porous media 1

1. POROSITY OF RESERVOIR ROCKS

11. Definitions and theory 7

12. Accuracy of porosity measurements 9

13. Direct methods for measuring rock porosity in the laboratory 11

13.1. Measurement of porosity for fresh or preserved samples known as the fluid summation method 12

13.2. Measurement of porosity for extracted or exposed sample . . .

13.21. Preparation of samples 12
13.22. Measurement of the total volume V_T 13
13.23. Measurement of solid volume V_S 15
13.24. Measurement of pore volume V_P 20
13.25. Measurement of total porosity 24

14. Cases of cracked, vesicular or macroporous media 25

15. Case of unconsolidated media 25

16. Determination of porosity by indirect methods (a posteriori on the basis of logs: electric-neutron-gamma/gamma logs) 26

16.1. Formation Factor: FF 26

16.2. Determination of ϕ by macroresistivity logs 27

16.3. Determination of ϕ by microresistivity logs 30

16.4. Determination of porosity by means of neutron logging 30

16.5. Determination of porosity by means of gamma-gamma logging 32

16.6. Determination of porosity by acoustic logging 32

17. Effect of compaction on porosity 34

18. **Comparison of the various methods utilized for porosity measurements** . . . 38

19. **Appendixes I and II** . . . 38
Appendix I. Mechanical properties of reservoir rocks . . . 38
Appendix II. Effect of compaction on pore volume . . . 41

2. PERMEABILITY OF RESERVOIR ROCKS

21. **Definition. Theory. Darcy's law** . . . 43
21.1. Definition of permeability and Darcy's law . . . 43
21.2. Units . . . 44
21.3. Slippage at the walls: Klinkenberg effect . . . 45
21.4. Specific, effective and relative permeabilities . . . 47

22. **Validity of Darcy's law** . . . 50
22.1. Real gas flows . . . 50
22.2. Liquid flows . . . 51
22.3. Attempt to establish a rate of flow law which is valid for both gas and liquids . . . 52
22.31. Integration of the quadratic equation in permanent flow: radial circular flow . . .
22.32. Parallel flow when integrating from 0 to 1 . . . 56
22.33. Quadratic equation for pressure drops along the path dr . . . 56
22.4. Conditions to be respected for assuring validity of Darcy's law. 57
22.41. Case of air flow . . . 57
22.42. Case of liquid flow . . . 58

23. **Average permeabilities for several layers** . . . 60
23.1. Parallel layers . . . 60
23.11. Parallel linear flow . . . 60
23.12. Radial circular flow . . . 61
23.2. Series of layers . . . 62
23.21. Parallel flow . . . 62
23.22. Radial circular flow . . . 63

24. **Variations in permeability in the case of fluid-medium reactions** . . . 64

25. **Effect of geostatic pressure on permeability** . . . 66

26. **Relation between permeability and pore morphology** . . . 67

27. **Evaluation factor for permeability with other parameters: concept of tortuosity, the coefficient of Kozeny** 69

28. **Physical analogies with Darcy's law** 71

28.1. Ohm's law 71

28.2. Fourier's equation for heat transmission 72

29. **Laboratory measurement of reservoir rock permeability with single-phase fluids, permeameter** 72

29.1. Measurement of air permeability k_a for geometric samples (with parallel flow) 73

29.11. Variable head permeameter, IFP type (derived from the instrument of M. Métrot 73

29.12. Constant head permeameter, Core Laboratories type 75

29.2. Measurement of gas permeability for hollow cylindrical samples: dry gas 76

29.3. Permeability measurements for large samples (air permeability) 77

29.4. Utilization of the Hassler assembly 77

29.5. Permeability of unconsolidated media 79

2.10. **Indirect determination of permeabilities and correlations between porosity and permeability** 81

2.10.1. Thin section porosity 81

2.10.2. Estimate of permeability by means of logs 82

2.10.3. Calculation of permeability on the basis of pore distribution 82

2.10.4. Correlation between internal surface areas of porous media and permeability 83

2.10.5. Permeability-porosity correlations 84

3. FLUID SATURATION OF RESERVOIR ROCKS AND CAPILLARY PROPERTIES

31. **Fluids contained in reservoir rocks and interstitial water** 85

31.1. Fluids contained in the reservoir 85

31.2. Interstitial water. Definition 86

32. **Capillary properties of rocks** 87

32.1. Review of some ideas concerning surface phenomena 87
32.11. Fluid/fluid interface and interfacial tension. Capillary pressure 87
32.12. Solid-interface contact and angle of contact 88
32.13. Factors affecting interfacial tensions between reservoir oil and water 90
32.14. Capillary pressure in a field. Definition.................. 90

32.2. Solid wettability. Threshold displacement pressure 92
32.21. Solids and wettable reservoirs........................ 93

32.3. Morphological study of pores by mercury injection......... 94
32.31. Method and equipment 94
32.32. Interpretation of capillary pressure curves by mercury injection 96
32.33. Utilization of capillary pressure curves by mercury injection: concept of the most frequent pore 96

33. Oil migration. Effects of capillary forces 97

33.1. Capillary migration 97

33.2. Final field equilibrium................................ 99

33.3. Effects of a pressure gradient.......................... 101

34. Interstitial water determination in the laboratory 101

34.1. Measurement of capillary pressure by the restored state method 103

34.2. Measurement of capillary pressure by mercury injection (Purcell method) 108

34.3. Measurement of capillary pressure by evaporation (Messer method)...... 110

34.4. Measurement of capillary pressure by centrifuging.......... 111

34.5. Comparison of different methods for determining interstitial water by capillary pressure measurements 111

34.6. Conversion of laboratory results 112

34.7. Average results for capillary pressures and correlations 113
34.71. The function of capillary pressure 114
34.72. Second correlation method for calculating connate water saturation 115
34.73. Correlation of irreducible water saturation and permeability-porosity 115

35. In situ determination of interstitial water saturation (with electric logs) 117

35.1. Basic situation 117

35.2. Case of a water-bearing zone 118

35.3. Determination by means of Formation Factor. **118**

35.4. Tixier method 118

35.5. Rapid simplified approach 119

36. Indirect methods for determining fluid saturation based upon rock analysis 120

36.1. Determination of gas saturation 121

36.2. Determination of water and oil saturation by distillation at atmospheric pressure 121

36.3. Determination of water and oil saturation by vacuum distillation 124

36.4. Determination of water content by solvent extraction (Dean-Stark distillation) 125

36.5. Determination of water content by reaction with calcium hybride 125

36.6. Determination of oil content by colorimetry 126

4. NATURAL ROCK RADIOACTIVITY MEASURED IN THE LABORATORY

41. Natural radioactivity of drill cuttings 127

42. Natural radioactivity of cores 129

43. Examples of the interpretation of natural radioactivity measured on cores 130

5. CORE ANALYSIS AND INTERPRETATION OF RESULTS

51. General remarks. Classification of core analysis 139

51.1. Core analysis 139

51.2. Special core analysis. 140

52. Conventional core analysis 140

52.1. Removal of cores 140

52.2. Sampling. Collection of samples and preservation 141

52.3. Conventional core analysis 142
52.31. Partial core analysis 142
52.32. Complete core analysis by summation of fluids 142
52.4. Side wall core analysis 146
52.41. Distillation method (retort method) (at atmospheric pressure) . . 146
52.42. Extraction with pentane (pentane cut method) 146
52.5. Whole core analysis 148

53. Interpretation of core analysis 150
53.1. General remarks 150
53.2. Basic considerations 151
53.3. Qualitative interpretation: relation between depth and permeability, porosity and lithology 151
53.4. Quantitative interpretation: distribution of fluids and different types of production 151
53.5. Examples of interpretation 152
53.51. FRIO formation: clean sands 152
53.52. COCKFIELD formation: relatively clean sands 153
53.53. Examples 154

54. Calculations 156

55. Applications of core analysis 160
55.1. Exploration 160
55.2. Well completion and workover operations 160
55.3. Field development 161
55.4. Well and reservoir evaluation 161

REFERENCES 163

SUBJECT INDEX 167

introduction

reservoir rock characteristics objective of core analysis and general presentation of porous media

Examination and analysis of cores are probably the most important basic technique available to the petroleum industry for obtaining optimum recovery in the exploitation of reserves.

Core analysis as we know it today practically did not exist at the beginning of petroleum prospecting, and drillers and producers considered their operations fully successful if the well began to flow during drilling. We now know, however, that many of the old wells have produced only a small portion of their oil reserves.

It was therefore reasonable that tests should be made for recovering and examining fragments of the oil-bearing rock. The first core analyses were an art and a question of characteristic smells and tastes, but these methods along with visual observation of lithological properties supplied interesting data to geologists, producers and builders of drilling equipment.

At the present time, there are many other techniques available to geologists and producers for completing a well and evaluating and determining oil and gas properties, etc. Core analysis is still essentially a basic tool for obtaining direct and interesting data concerning the rocks drilled.

Continuous mechanical coring should be utilized only with care since :

(a) It is more expensive than normal drilling.
(b) Recovery of the rocks drilled is not complete.

Diamond core drilling has, however, made it possible to obtain costs very close to those of normal drilling and recovery rates which are close to 100%.

In addition, some core barrels are capable of making excellent recoveries in unconsolidated formations.

The cores taken during drilling constitute samples of the reservoir rocks containing the fluids sought after in gaseous or liquid form. These reservoir rocks are porous media whose intergranular nature is fundamental to their definition and properties. A porous medium can have the following properties:

(a) It may be consolidated or unconsolidated.

(b) It may be clean (without clays) or it may contain clays, and cements which bind the grains together and are capable of reacting with the fluids present. It may also be silicified.

(c) A porous medium may be homogeneous or fissured.

(d) The high specific surface area of this medium will influence absorption phenomena and solid/fluid contacts.

(e) A porous medium has certain affinities with the fluids is contains.

The distribution of the fluids coexisting in the space depends upon:

. Structural shapes.
. Rock and fluid composition.
. Pressure and temperature of the reservoir.
. The movement preceding fluid equilibrium (since it is always difficult to observe equilibrium in a porous medium).

(f) A porous medium is also characterized by the existence of an aqueous phase. It was established that interstitial water exists in all hydrocarbon accumulations.

The properties of a porous medium (reservoir rock) should be determined according to a certain order:

(a) Storage capacity (porosity).
(b) Circulation capacity (permeability).
(c) Capillary pressures and saturations.
(d) Correlations: porosity/permeability, saturation/permeability, etc.
(e) Lithological coefficients.
(f) Electrical properties.
(g) New measurements such as determination of saturation by X rays, etc.
(h) Statistical treatments and averages.
(i) Estimate of reserves.

The samples are described first.

This description is very often accompanied by photographs (preferably in color).

Hydrocarbon reservoirs consisting of porous media show:

(a) Physical (or petrophysical) properties.
(b) Mechanical properties.
(c) Electrical properties.
(d) Surface properties.

Planning for production from a hydrocarbon field requires that the following be known as soon as possible:

(a) The quantity of hydrocarbons present (estimate of reserves).
(b) Properties of the reservoir (storage) in the various parts.

These two objectives can be attained:

1. By collecting a sufficient number of representative samples with good spatial distribution.
2. By analysing these samples and correctly interpreting the results of this analysis.

Continuous coring cannot be systematically utilized in all wells, but an attempt should be made to have complete cores for some intervals which will make it possible to verify the interpretations. It should be noted that some properties, such as permeability, can really be determined only by means of core measurements.

Core drilling is all the more necessary as reservoir properties show greater variations. Probability laws must be determined for the various values of *in situ* porosity, permeability and saturation (the physical properties of hydrocarbon reservoirs are the subject of the following chapters). These properties vary spatially, i.e. anisotropy, heterogeneity.

As to estimates of reserves, conventional statistics can only approximately give the level of error involved in the evaluation since the method takes into account only the number of wells and not their location. On the other hand, geostatistics makes it possible to solve the problem of the weighting of the various wells and to calculate an evaluation error depending upon the spatial continuity of the variable, possible errors in measurement and the positioning of the well in the field.

Core measurements are intended to achieve several goals and are distinguished by their order of urgency and whether it is a question of an exploratory or a development well:

(a) Case of an exploratory well (wildcat): core analysis makes it possible to recognize the structure, determine its physical characteristics and estimate production possibilities.
(b) Definition of optimum completion: core measurements make possible:
- Selection of intervals for testing.
- Interpretation of drill stem tests.
- Determination of the best combinations for the order of completions if there are several horizons.
- Evaluating effectiveness of completion.

(c) Development of fields and estimates of reserves. For example, by means of core analysis it is possible to:

. Defining limits of field.
. Determine contacts between the various fluids, and variations across field.
. Establish structural and stratigraphic correlations.
. Determine net pay.
. Estimate reserves and initial production.

Data furnished by core analysis will be completed and compared or correlated with those furnished by the various types of logging, either mud logging (done during drilling) or after drilling has been completed such as electrical and nuclear logging. Some core measurements can serve as sampling and as a basis of interpretation for other logging.

Laboratory analysis of cores will determine the physical properties of reservoir rocks such as porosity, permeability and pore morphology and will make it possible to determine the nature of the fluids contained in the rock (water, gas, oil) as well as the amounts of these fluids present.

The natural radioactivity of the formations, which is also a physical property of reservoir rocks, can also be measured easily without destruction, loss or alteration of the core, and the results obtained may be useful for correlations or for locating losses in cases of core recovery rates under 100%.

Classification in a core analysis can be carried out in various ways as follows:

(a) Sample or core size.

(b) Core condition: fresh or preserved, exposed or extracted, consolidated or unconsolidated.

(c) Nature and quantity of required data.

The following are distinguished:

1. **Conventional core analysis,** consisting of:

(a) For fresh or preserved samples (complete results):

. Porosity.
. Permeability.
. Gas, water, oil saturation.

(b) For weathered or extracted samples (partial results, no saturation measurements):

. Porosity.
. Permeability.

This conventional analysis can be carried out on samples of very different sizes, i.e.:

(a) On small samples (little plugs) collected at regular intervals: conventional core analysis proper.

(b) On whole cores, especially in the case of fissured deposits: whole core analysis.

(c) On samples from side wall cores: side wall core analysis.

Methods vary for each case or require special equipment, but the data obtained are those indicated above making it possible to progressively plot a diagram which is very useful for the study of the reservoir and the continuation of drilling.

2. **Special core analysis.** While in the above-mentioned cases all samples are studied and analysed, in the case of special core analysis measurements are made on a limited number of samples, whether fresh or not, which are selected in terms of their physical properties, porosity or permeability. These measurements, which are necessary for the study of reservoir problems, consist of:

(a) The study of capillary pressures, for example, either by means of the restored-state method or mercury injection.

(b) Measurements of the formation factor and resistivity ratio.

(c) Two-phase flows and the study of relative permeabilities:

gas/oil ; water/oil ; gas/water

(d) Study of wettability tests and various tests depending upon the problems to be solved.

Paragraphs (b), (c) and (d) are not treated in this course, but solely paragraph (a) concerning capillary pressures measured by two different methods, and pore morphology by means of mercury injection will be dealt with.

1

porosity of reservoir rocks

11. DEFINITIONS AND THEORY

Sedimentary rocks consist of grains of solid matter with varying shapes which are more or less cemented, and between which there are empty spaces (Fig. 11.1). It is these empty spaces which are able to contain fluids such as water or liquid or gaseous hydrocarbons and to allow them to circulate. In this case the rock is called porous and permeable.

Let us consider a rock sample of any shape (whether geometrical or not):

V_T = total (or apparent) volume of the sample,
V_P = volume of hollow spaces (or pore volume) between the solid grains,
V_S = real volume of grains.

The porosity of the sample is the ratio of V_P to V_T:

$$\text{Porosity} = \frac{\text{pore volume}}{\text{total volume}} = \frac{\text{total volume} - \text{solid volume}}{\text{total volume}}$$

Porosity is generally designated by the letters ϕ or f.
It is expressed as a percentage or fraction:

$$\phi\% = \frac{V_P}{V_T} \cdot 100 = \frac{V_T - V_S}{V_T} \cdot 100 \qquad \text{(Eq. 11.1)}$$

The **effective porosity** of the sample is the ratio of the volume of interconnected pores to the total volume of the sample.

Pores may exist which are not interconnected. **Residual porosity** can therefore be defined as resulting from those pores which are not interconnected. The total porosity (effective + residual) is established by electrical or nuclear logging or *a posteriori.*

Porosity varies with:

(a) Grain size and shape.

(b) Grain distribution.

Thus

(a) A cubic arrangement (Fig. 11.2) of spherical grains gives $\phi = 47.6\%$ (maximum).

(b) A rhombohedral arrangement (Fig. 11.3) of spherical grains give $\phi = 26\%$.

(c) A cubic arrangement of spherical grains of two different sizes (Fig. 11.4) gives $\phi \approx 12.5\%$.

Some real values for measured porosities are:

(a) Sandstones: 10 to 40% depending upon the nature of the cement and their state of consolidation.

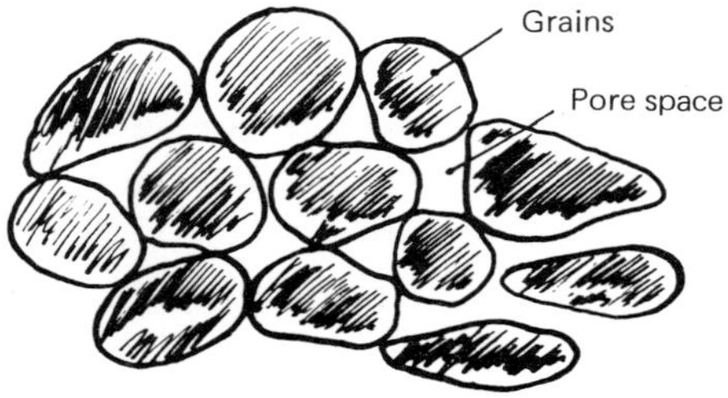

Fig. 11.1. Porous media.

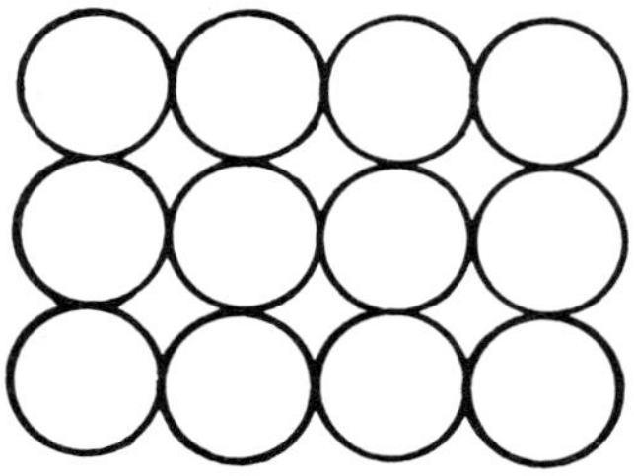

Fig. 11.2. Cubic packing. (Equal grain size) $\phi = 47.6\%$.

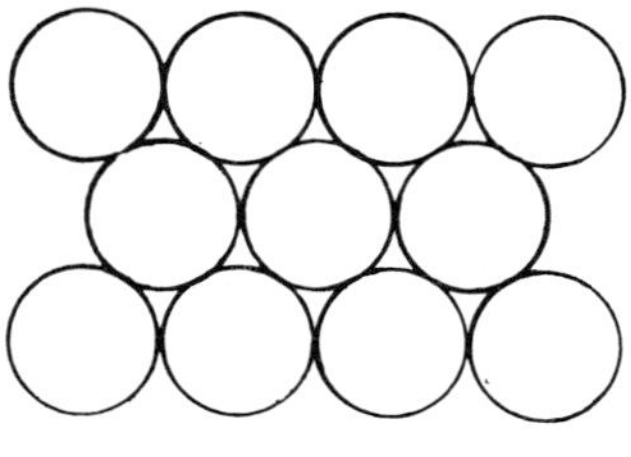

Fig. 11.3. Rhombohedral packing. (Equal grain size) $\phi \approx 26\%$.

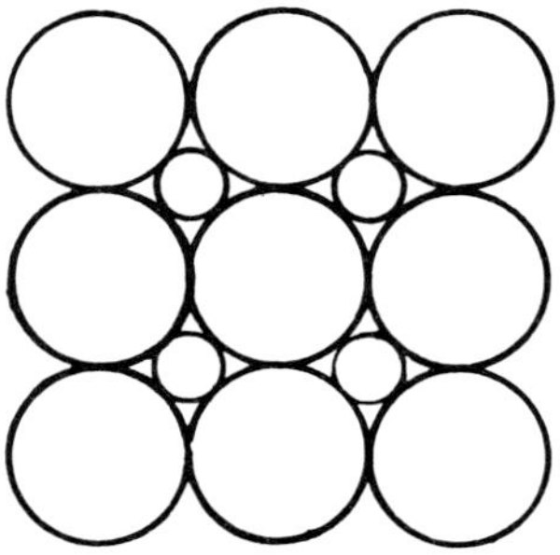

Fig. 11.4. Cubic packing. Two grain sizes $\phi \approx 12.5\%$.

(b) Limestones and dolomites: 5 to 25%.
(c) Clays: 20 to 45% depending upon origin and depth.

It is generally said that porosity is:

(a) Negligible if $\phi < 5\%$.
(b) Low if $5 < \phi < 10\%$ (e.g. Hassi-Messaoud).
(c) Good if $10 < \phi < 20\%$ (e.g. Zarzaïtine, Cazaux, etc.).
(d) Very good if $\phi > 20\%$ (e.g. Chateaurenard).

Examples of pores and cores

(a) Relatively consolidated sand or sandstone (Fig. 11.5). The grains are relatively angular.

(b) Limestone (or carbonate reservoir). Their porosity is very variable. The intercrystalline porosity is equivalent to that of the sandstone mentioned above but is much lower (pore sizes are much smaller). These rocks can show relatively large vesicles which therefore increase porosity, or cracks (Fig. 11.6). The latter cause a very small rise in porosity but, on the other hand, increase permeability considerably.

(c) Examples of porous reservoir rocks (see Fig. 11.7).

12. ACCURACY OF POROSITY MEASUREMENTS

The calculation of error below shows that it is difficult to measure porosities with low relative error. In effect we have:

$$\phi = \frac{V_P}{V_T}$$

and

$$\frac{\Delta\phi}{\phi} = \frac{\Delta V_P}{V_P} + \frac{\Delta V_T}{V_T} \qquad \text{(Eq. 12.1)}$$

(variation in absolute value)

Assuming that V_T and V_P are determined with the same absolute error ΔV(1) and replacing V_P by its equivalent ϕV_T, we have:

$$\frac{\Delta\phi}{\phi} = \frac{\Delta V}{\Delta T}\left[1 + \frac{1}{\phi}\right] \qquad \text{(Eq. 12.2)}$$

(1) This is not the case for all the methods utilized.

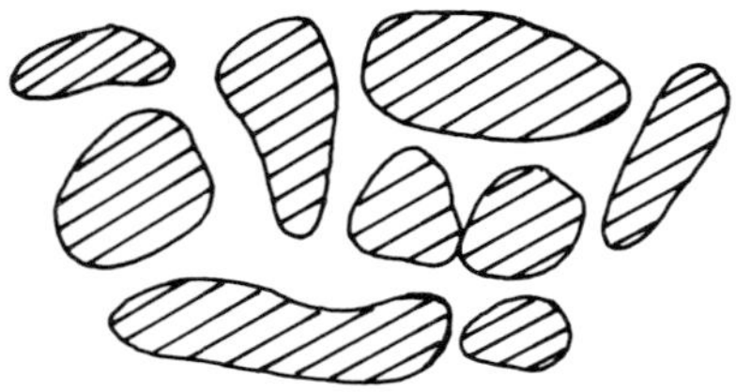

Fig. 11.5. Characteristic sand.

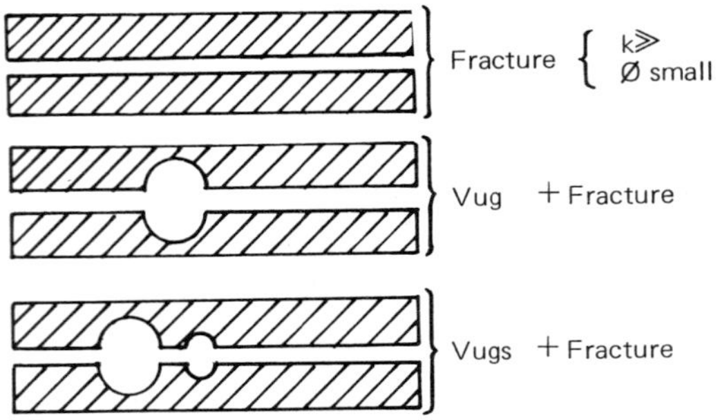

Fig. 11.6. Idealized limestone with fractures and vugs.

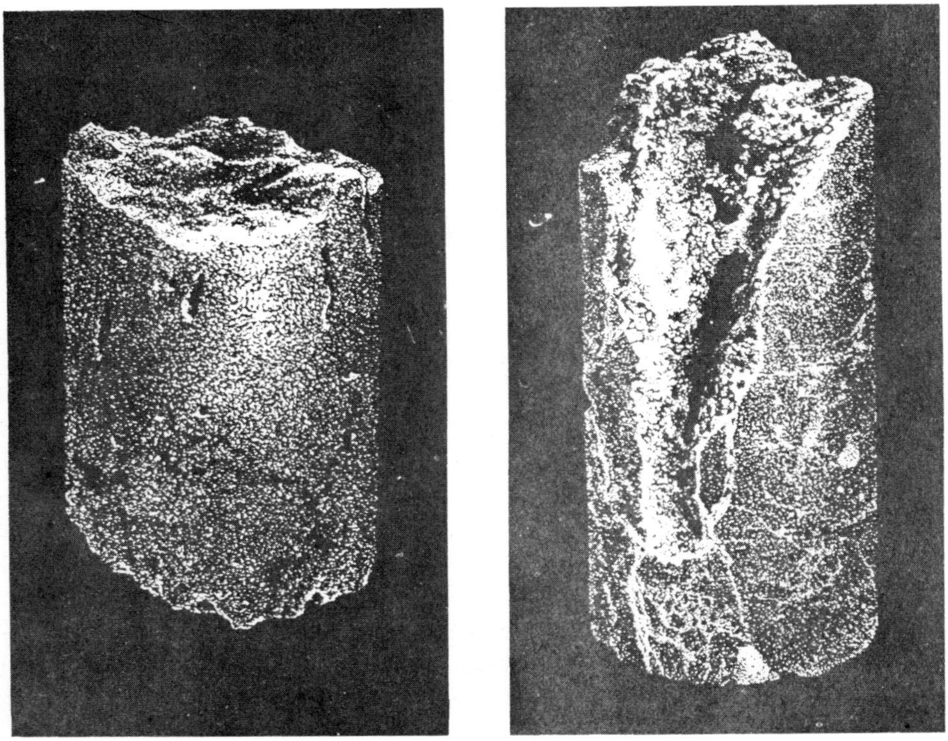

Fig. 11.7.

If the porosities are very small, i.e. if ϕ is between 1 and 5%, it is possible to write:

$$\Delta\phi \# \frac{\Delta V}{V_T} \qquad \text{(Eq. 12.3)}$$

if $V_T = 10 \pm 0.01\ \text{cm}^3$, we have:

$$\Delta\phi = 0.1\%$$

i.e. if $\phi = 1\%, 0.9\% < \phi < 1.1\%$.

It is therefore indispensable to have accurate instruments available and to utilize them carefully so as to measure low porosities with acceptable accuracy.

Since the error is inversely proportional to the total sample volume, it is recommended that the samples be as large as possible.

TABLE OF RELATIVE ERRORS FOR POROSITY (IN %) FOR DIFFERENT VALUES OF V_T
(ON THE HYPOTHESIS OF $\Delta V = 0.01\ \text{cm}^3$)

V_T(cm^3) \ ϕ (%)	1	2	4	8	12	16	20	25
2	50.5	25.5	13.0	6.75	4.25	3.6	3.0	2.5
4	25.2	12.7	6.5	3.4	2.1	1.8	1.50	1.25
8	12.6	6.4	3.25	1.7	1.1	0.9	0.75	0.6
16	6.3	3.2	1.6	0.85	0.5	0.45	0.37	0.3
32	3.0	1.6	0.8	0.4	0.26	0.23	0.18	0.15
64	1.6	0.80	0.4	0.2	0.13	0.11	0.094	0.08

13. DIRECT METHODS FOR MEASURING ROCK POROSITY IN THE LABORATORY

Direct methods for measuring porosity differ depending upon the nature of the sample and its size:

(a) Fresh or preserved samples (fluid summation method).
(b) Exposed or extracted samples.

13.1. Measurement of porosity for fresh or preserved samples known as the fluid summation method

This method was developed by *Core Laboratories Inc.* (US Patents 2 282. 654; 2 345. 535 and 2 361. 844) and will be presented in detail in Chapter 5 of this course. The principle is as follows:

The **fresh** soil sample contains the following fluids: gas, water, oil. The quantities of the various fluids present are determined in % of V_T, and the porosity is their sum.

The gas content is determined by mercury injection.

Water and oil are determined by distillation at atmospheric pressure (Retort Method).

As we shall see in Chapter 3, saturations are calculated on the basis of the above results.

This method, which is excellent for the analysis of fresh cores, operates on large quantities of cores, i.e. 150 to 250 g. The liquids collected are measured directly. Gaging is necessary for the oil, and the temperature should be controlled so as to avoid dehydration of the clays. The presence of salt has practically no effect.

13.2. Measurement of porosity for extracted or exposed samples

In order to determine:

$$\phi = \frac{V_P}{V_T} = \frac{V_T - V_S}{V_T}$$

it is necessary to measure 2 of the 3 factors V_T, V_S or V_P.

13.21. Preparation of samples

All the techniques described below, except for the fluid summation method, require samples from which the fluids have been extracted for the measurement of porosity. The samples may have a geometric shape (cube or cylinder), $V_T = 5$ to $70\ cm^3$, or any other shape, in so far as possible without sharp edges.

Extraction and washing can be carried out by:

(a) Soxhlet type washer extractors (Fig. 36.41).
(b) Dean-Stark type extractors.
(c) Vacuum retorting.
(d) Centrifuging and washing.

The method combining a Soxhlet and a centrifuge is very rapid and effective in conventional rock analysis. Vacuum retorting is no doubt the best solution for large cores except for clayey samples.

Amongst the most generally used solvents should be mentioned toluene, xylene, chloroform, carbon tetrachloride, acetone, Chlorothene, hexane, etc.

Drying is very important following extraction, especially if minerals which can be dehydrated are present. In some cases vacuum drying at low temperatures is possible.

In the laboratory it is indispensable to standardize the various operations for extracting, washing, drying and storing the samples before the various measurements are made.

13.22. Measurement of the total volume V_T

13.221

This is determined by measuring if the sample has a simple geometrical shape (cube or cylinder) and is not chipped or notched. A sliding caliper is used.

13.222

Determination by means of *a mercury volume pump* (Fig. 13.222.1). Before inserting the sample, the mercury is brought to a fixed mark above the sample chamber. The pump is brought to zero by means of a calibrated disk. The piston is removed and the sample is placed in the chamber, and the mercury is then brought back to the previously mentioned mark. The total volume is given by a reading of the pump gage.

Measurement accuracy : $\Delta V = \pm 0.01$ cm. Measurement is performed rapidly. The method is valid only if the mercury does not enter the pores of the sample.

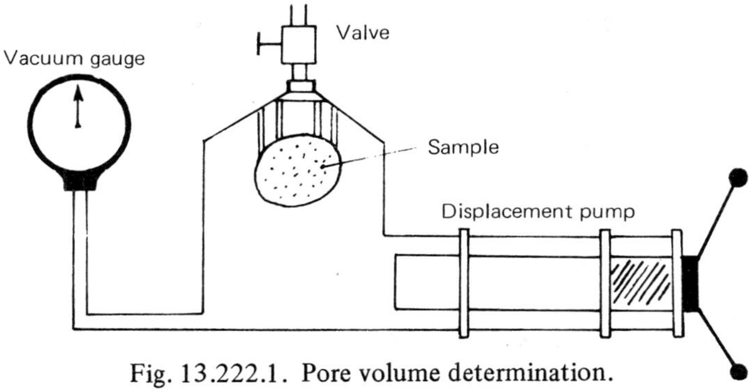

Fig. 13.222.1. Pore volume determination.

The base of the core immersed in the mercury can be located between 3 and 6 cm from the free mercury level. For a distance of 5 cm, the radius of the pores invaded by the mercury would be greater than $r_c = 0.014$ cm.

13.223

Determination by *measurement of the buoyancy* acting on a sample immersed in mercury "voltomètre" of the *Institut Français du Pétrole*).

Figure 13.223.1 shows the casing C with the float F mark point R and balance scale B. The core is placed on the balance and weights are added until there is contact between the mark point and the mercury. The core weight C, the casing weight P_c, buoyancy affecting the immersed part P_a, the weights P_1 and the surface forces f_s ensure equilibrium according to the equation:

$$C + P_c + P_1 = f_s + P_a \qquad \text{(Eq. 13.223.1.)}$$

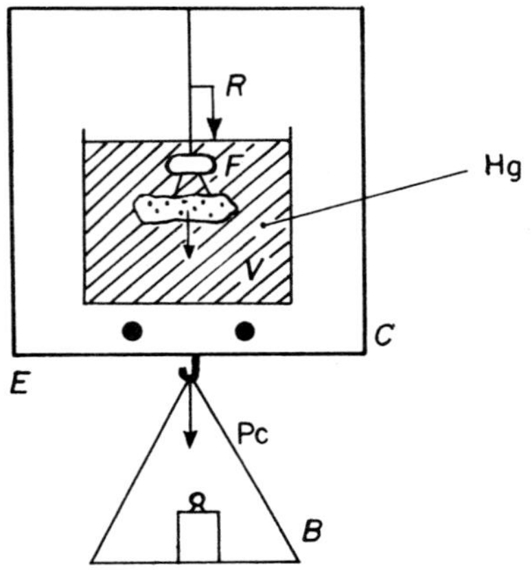

Fig. 13.223. IFP "voltomètre".

The core is then placed in the mercury under the float and equilibrium is reestablished with weight P_2. Taking into account the effects of buoyancy P on the core, we have:

$$C + P_c + P_2 = f_s + P_a + P \qquad \text{(Eq. 13.223.2.)}$$

If we subtract Eq. 13.223.1. from Eq. 13.223.2. we have:

$$P = P_2 - P_1$$

and therefore

$$V_T = \frac{P}{\rho_{Hg}} = \frac{P_2 - P_1}{\rho_{Hg}} \qquad \text{(Eq. 13.223.3.)}$$

ρ_{Hg} is the specific gravity of mercury under measurement temperature conditions.

It should be noted that this method is valid only if the mercury does not enter the pores. The immersion conditions for the sample are the same as in 13.222.

13.23. Measurement of solid volume V_S

V_S can be determined by the following three methods:

(a) By means of a picnometer.

(b) Immersion method: measurement of buoyancy for the solid phase in a wetting fluid.

(c) Utilization of a pressure chamber.

The first two procedures suppose that the sample is saturated. The third is based upon the Mariotte-Boyle law.

The following sequence of operations is necessary for saturating a porous sample (or many simultaneously) by means of a wetting fluid (Figs. 13.231 and 13.232):

(a) High vacuum with dry samples.

(b) Vacuum eliminated by a gas which is soluble in the solvent utilized.

(c) High vacuum.

(d) Saturation of samples under vacuum with a degasified solvent.

(e) Establishment of atmospheric pressure and placing of saturated samples under pressure of the order of 150-200 bars. The time necessary for placing the saturated samples under pressure depends upon their permeability, and this period becomes longer as permeability falls. The laboratory saturation operations should be standardized.

13.231. Use of a picnometer

Supposing that

P_0 = the weight of the empty picnometer,
P_1 = the weight of the picnometer full of a solvent with density ρ,
V = the volume of the picnometer.

we then have:

$$P_1 = P_0 + \rho V \qquad \text{(Eq. 13.231.1.)}$$

If the picnometer containing the saturated sample and filled with solvent is weighed again we have:

$$P_2 = P_0 + P_S + \rho(V - V_S) \qquad \text{(Eq. 13.231.2.)}$$

in which

P_2 = new weight of the picnometer with sample,
P_S = weight of the dry sample,
V_S = volume of the sample.

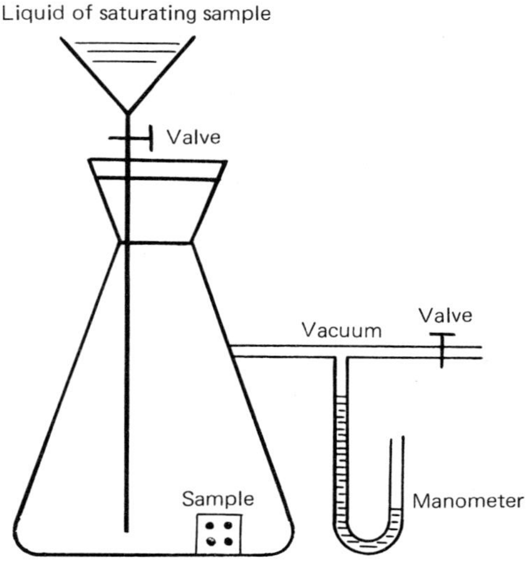

Fig. 13.231. Vacuum saturation of samples.

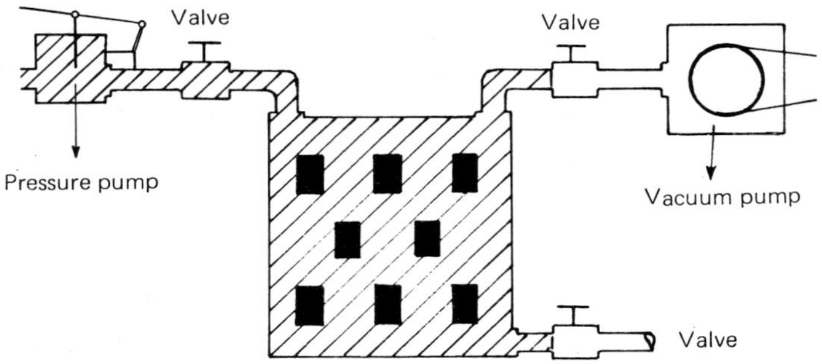

Fig. 13.232. Sample saturation in vacuum under pressure.

If we subtract (Eq. 13.231.1) from (Eq. 13.231.2) we have:

$$P_2 - P_1 = P_S - \rho V_S \qquad \text{(Eq. 13.231.3.)}$$

and therefore

$$V_S = \frac{P_S - P_2 + P_1}{\rho} \qquad \text{(Eq. 13.231.4.)}$$

It can be seen that:

$$\Delta V_S = \frac{3\Delta P}{\rho}$$

On the supposition that the error for ρ is negligible, it is necessary that: $\Delta P < 0.005$ g with $\rho \approx 1.6$ for $\Delta V_S < 0.01$ cm^3.

13.232. Immersion method

V_S is determined by the buoyancy thrust on the solid phase of the sample immersed in a solvent (Fig. 13.232.1.).

The dry weight P_S is determined after washing and drying.

As was indicated above, the samples are saturated in a solvent with a density ρ at measurement temperature.

The immersed weight P_i is determined.

The calculation is made. Solid volume:

$$V_S = \frac{P_S - P_i}{\rho} \qquad \text{(Eq. 13.232.1.)}$$

It can be seen that:

$$\Delta V_S = \frac{2\Delta P}{\rho}$$

For $\Delta V_S < 0.01$ cm^3, it is necessary that $\Delta P < 0{,}008\ g$ with $\rho \approx 1.6$.

The solvents used have high densities so as to make the phenomenon more apparent. Generally, carbon tetrachloride (CCl_4: $\rho_{20°C} = 1.6$) or Chlorothene NU (1) ($\rho_{20°C} \approx 1.32$) are used. The latter has a lower specific gravity and is much less toxic than the former.

Toxicity of carbon tetrachloride:

(a) Maximum admissible concentration: 25 ppm.

(b) Minimum concentration detectable by smell: 80 ppm.

Precautions: ventilated area, frequent washing of hands after contact, removal of contaminated objects.

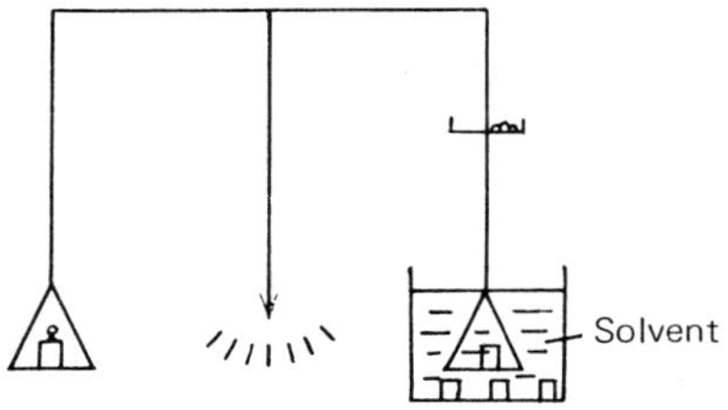

Fig. 13.232.1. Grain volume determination.

(1) Chlorothene or trichloroethane – 111 $CH_3 - C - Cl_3$

Remark:

The dry weight, P_S, and the solid volume of the sample, V_S, were determined in both methods and the real density can therefore be determined:

$$m_v = \frac{P_S}{V_S}$$

as well as the apparent density:

$$m_a = \frac{P_S}{V_T}$$

TABLE OF SOME GRAIN DENSITY VALUES
(in g/cm³)

Rock	$m_v = \frac{P_s}{V_s}$
Anhydrite	2.90 to 3.00
Clay	2.64 to 2.66
Limestone-Calcite	2.70 to 2.76
Dolomite	2.82 to 2.87
Feldspar	2.64 to 2.66
Sandstone	2.65 to 2.67
Gypsum	2.30 to 2.40
Quartz	2.59 to 2.66
Schist	2.64 to 2.66
Massive salt	2.14

13.233. Utilization of a compression chamber

Many apparatuses have been built in the United States and in Europe. They are based upon the Mariotte-Boyle law and exist for large or small cores. All (Fig. 13.233.1) include a chamber in which the sample is placed at a pressure which is changed with the variations in volume. The degree to which solid matter fills the chamber is then deduced.

There is a core holder with the volume V. When it contains a core of volume V_S, it does not contain more than $V - V_S$ of gas. If this gas is brought isothermally from P_1 to P_2, we have:

$$(V - V_S) P_1 = (V - V_S + \Delta V) P_2 \qquad \text{(Eq. 13.233.1)}$$

or

$$(V - V_S)(P_1 - P_2) = P_2 \, \Delta V \qquad \text{(Eq. 13.233.2)}$$

so that

$$V_S = V - \frac{P_2}{P_1 - P_2} \cdot \Delta V \qquad \text{(Eq. 13.233.3)}$$

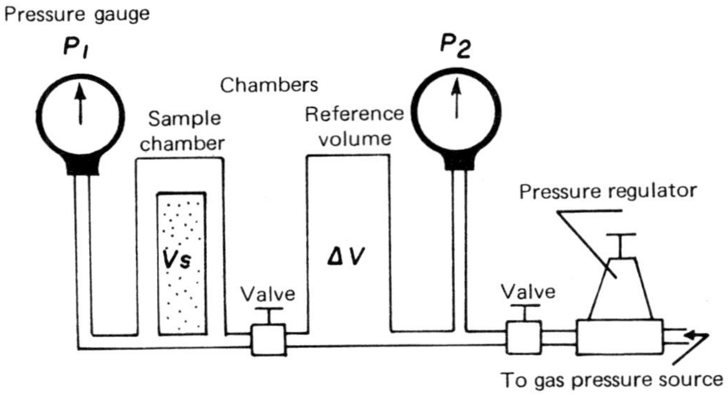

Fig. 13.233.1. Boyle's law porosimeter: grain volume determination.

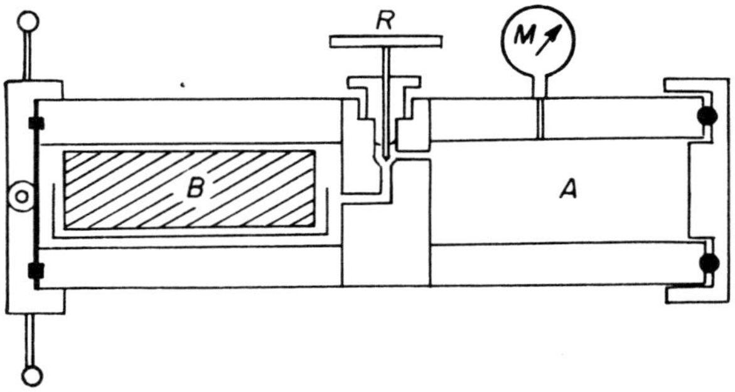

Fig. 13.233.2. Boyle's law grain volume determination (porosimeter).

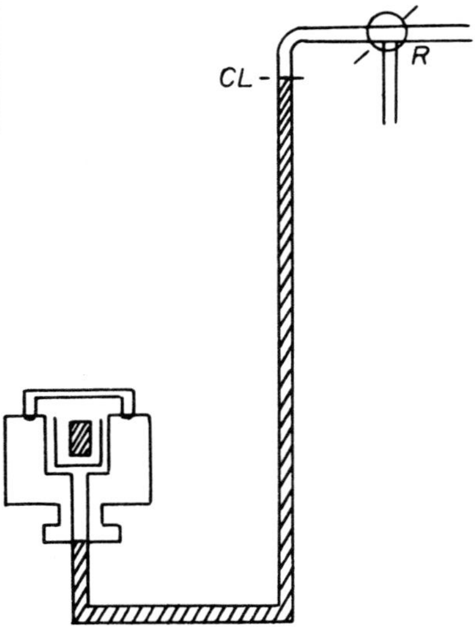

Fig. 13.233.3. Boyle's law porosimeter: pore volume determination.

Figures 13.233.1 and 13.233.2 show an apparatus for large cores, while the apparatus in Fig. 13.233.3 is for small samples.

These simple devices are subject to various errors which should not be overlooked:

(a) Readings are linked to atmospheric pressure.

(b) Adsorption of water vapor or of the gas utilized by clays can falsify the measurements to a high degree.

13.24. Measurement of pore volume V_P

Pore volume can be measured directly:

By measuring the air volume contained in the pores.
Or by weighing a liquid filling the pores.
Or by mercury injection.

13.241

Measurement of air contained in the pores, by means of the E. Vellinger type porosimeter or the Washburn Bunting porosimeter (Fig. 13.241.1).

The principle consists in isolating the air contained in the sample pores by means of a partial vacuum and then measuring it.

In principle the method is a good one since only the adsorbed air can falsify the results. In practice, it is difficult to manufacture these glass apparatus because of their fragility and the possibility of leaks, but French firms have used them very frequently.

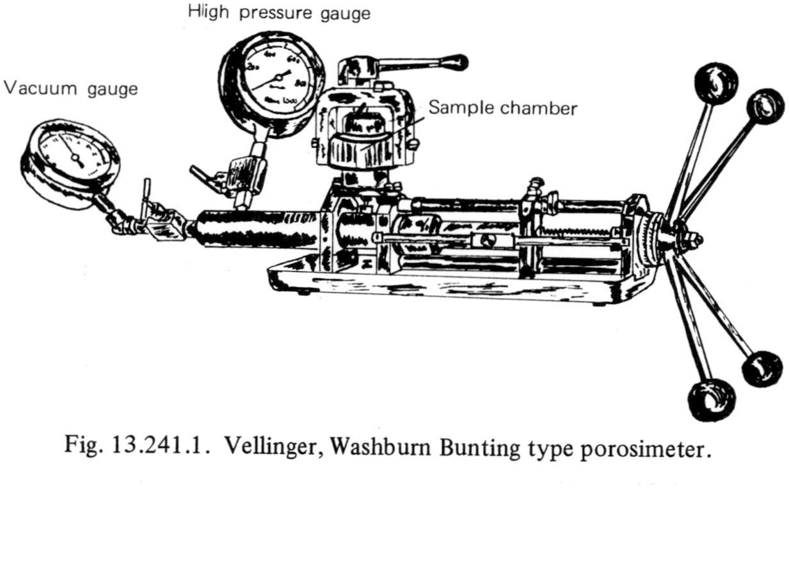

Fig. 13.241.1. Vellinger, Washburn Bunting type porosimeter.

13.242

Measurement of air contained in pores through release by means of a mercury volume pump: gas expansion method.

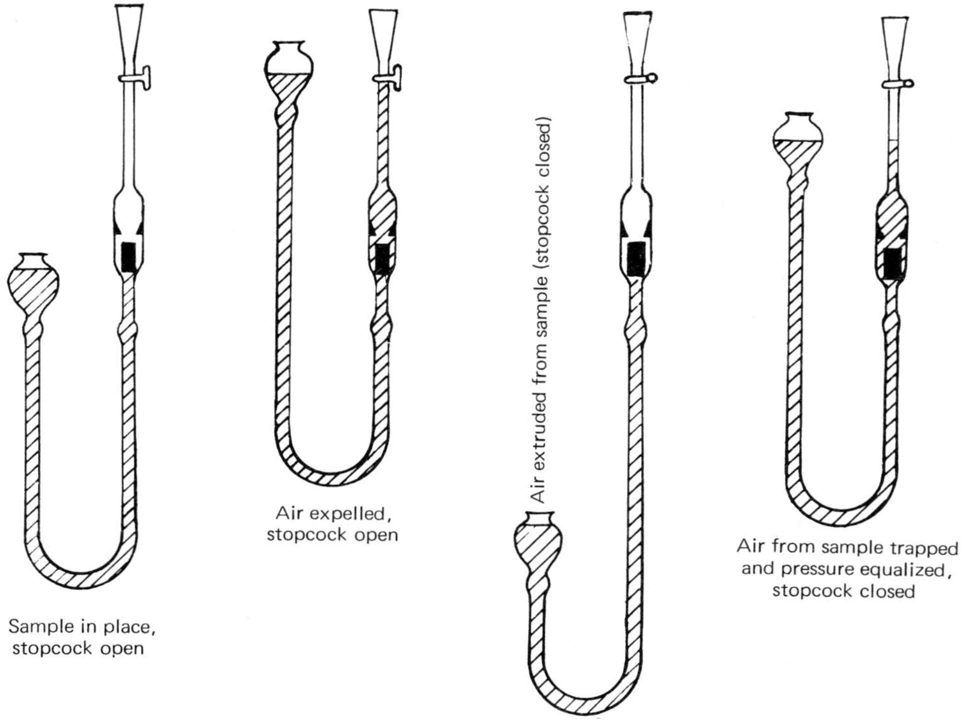

Fig. 13.242.1. Boyle's law porosimeter (mercury pump).

Principle of the method: The sample is placed in the sample chamber of the mercury pump which has a vacuum gage (Fig. 13.242.1). V_T can be determined immediately. The chamber is closed and the pump is brought to zero. Mercury is then withdrawn until the vacuum gage indicates a fall in pressure:

$$\Delta P = \frac{1}{2} \text{ atm} = \frac{H_0}{2}$$

We have (the Mariotte-Boyle law):

$$V_P \, . \, H_0 = (V_P + v)(H_0 - \Delta P) \qquad \text{(Eq. 13.242.1)}$$

Since

$$\Delta P = \frac{H_0}{2} = \Delta H_0$$

it follows that:

$$\nu = V_P$$

and

$$\phi = \frac{V_P}{V_T} \cdot 100 \qquad \text{(Eq. 13.242.2)}$$

Pore volume is equal to the volume of mercury withdrawn in order to make the pressure drop by one half atmosphere.

In reality atmospheric pressure is not always 76 cm of mercury, and corrections of the pump are necessary because of trapped air, etc.

In practice, operations take place at:

$$\frac{H_0}{\Delta H_0} = \text{Const.}$$

and

$$\frac{\Delta V_P}{V_P} = \frac{\Delta H_0}{H_0} = \frac{1}{f}, f = \frac{\text{pore volume}}{\text{pump reading}} = \frac{V_P}{\Delta V}$$

The coefficient f depends upon the volume of air contained in the pores and on the ΔP adopted which is close to one half atmosphere. It is sufficient to expand known volumes of air (0.5-1-2-3-4...N cm^3) at ΔP = Const. without a sample and to calculate $f\left(\text{close to 1 for } \Delta P = \frac{H_0}{2}\right)$.

For a given pump reading, f is read on the curve (Fig. 13.242.2) so that pore volume = fx reading.

(In practice, this sampling is carried out two or three times per day.)

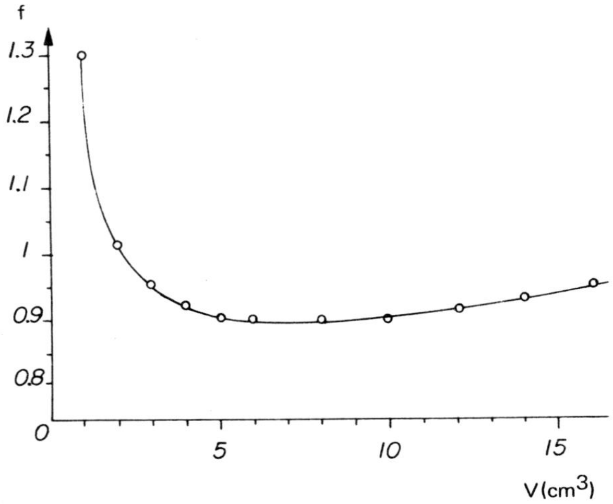

Fig. 13.242.2. Value of f versus air volume.

13.243. Measurement of pore volume by weighing the liquid filling the pores (Fig. 13.243.1)

The sample is extracted and dried and then saturated with brine as was indicated in paragraph 13.23 "Determination of V_S". In this case, the soluble gas in the liquid is CO_2. It is necessary that saturation be carried as far as possible and

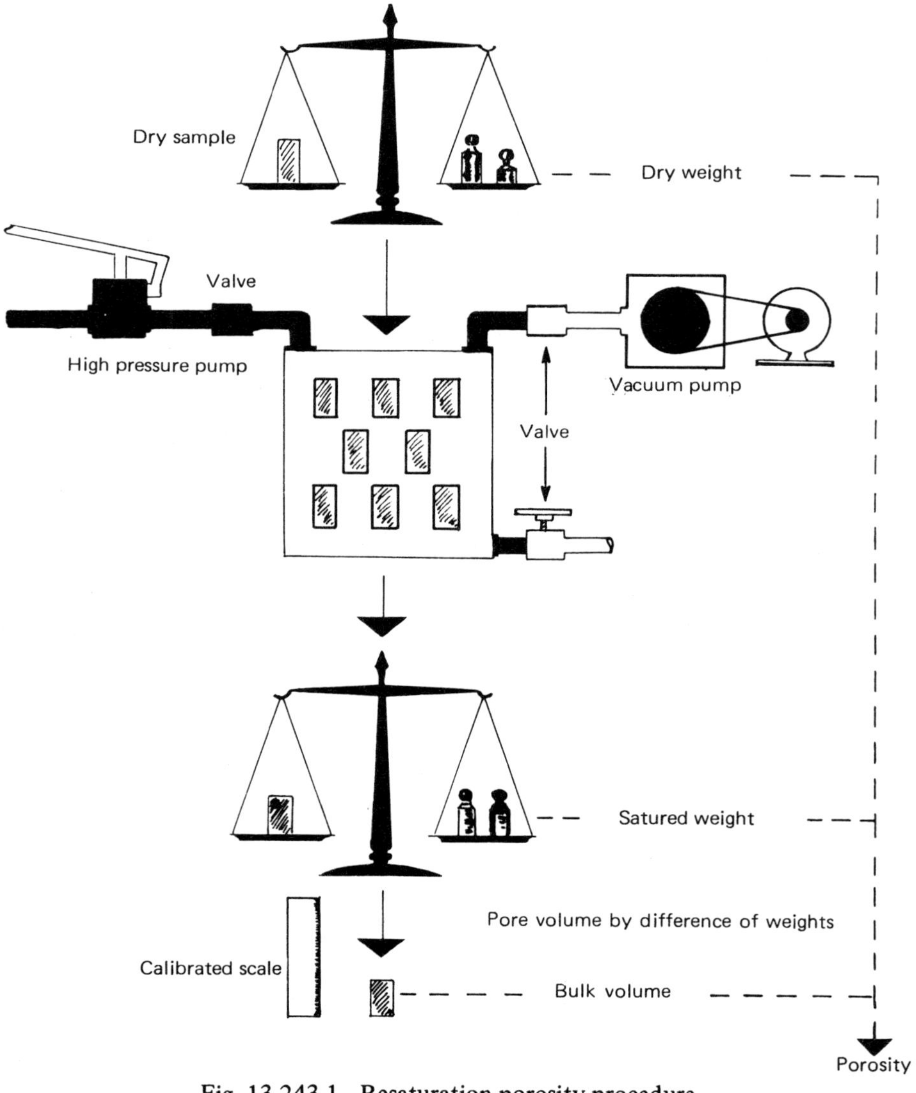

Fig. 13.243.1. Resaturation porosity procedure.

also that drops of parasitical water not be left on the surface of the sample. Blotters or cloth with excessively high capillary suction should not be used.

An analogous method was developed for drill cuttings by Westbrook, Redmond and Manesterski and utilizes a semi-permeable wall.

13.244. Measurement of pore volume by mercury injection

The principle consists of forcing mercury under relatively high pressure into the rock pores. An apparatus of the Ruska type is shown in Fig. 13.244.1.

The volume pump gives, firstly, V_T and then the volume of mercury injected in terms of pressure (Fig. 13.244.2).

The curve (Fig. 13.244.2) concerns a sample of sandstone where $\phi = 7\,\%$. It is clear that the volume of pores which is invaded depends upon the injection pressure. At 100 bars, partial porosity is 5.6%. In order to obtain $\phi = 7\%$ it is necessary to reach 250 bars.

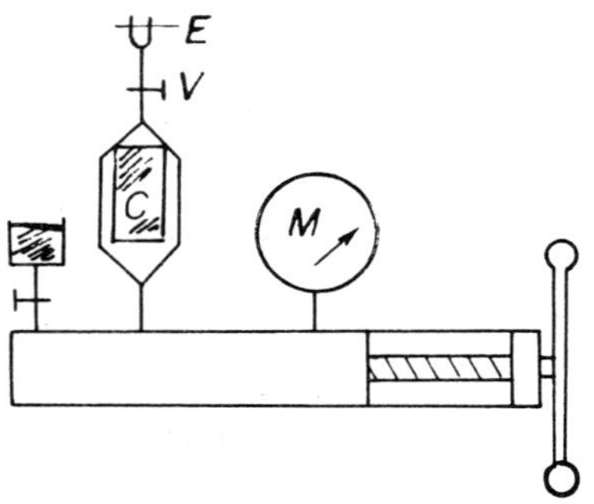

Fig. 13.244.1. Volume pump for mercury injection.

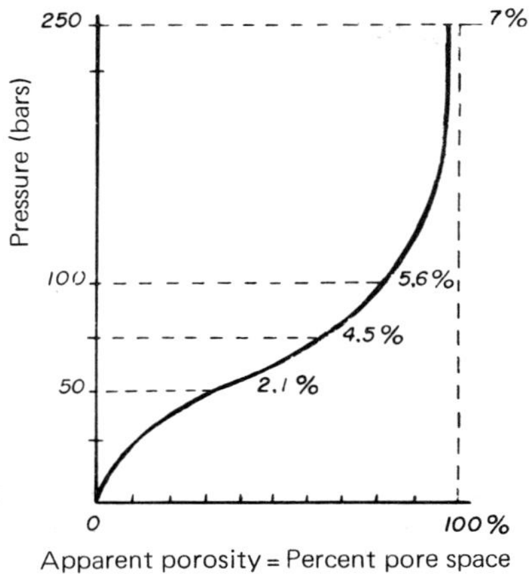

Fig. 13.244.2. Porosity through mercury injection.

13.25. Measurement of total porosity

The total volume is determined by one of the preceding methods. The sample is then ground very fine and the solid volume is determined:

(a) Either by utilizing a compression chamber.
(b) Or by liquid displacement.

If the dry weight is known the real and apparent densities can be calculated. The method is long and difficult but it gives the total porosity.

14. CASES OF CRACKED, VESICULAR OR MACROPOROUS MEDIA

In such cases only the analysis of large cores is valid.

The total volume was measured as has already been indicated. If pore volume is determined by the immersion method (paragr. 13.232), no distinction will be made between the porosities of macropores due to vesicles, geodes, etc. and matrix porosity as in the method for determining pore volume by mercury injection (paragr. 13.244).

Figure 14.1 shows mercury injection in a homogeneous sample with only matrix porosity, while Fig. 14.2 indicates the phenomenon for a sample with vesicles.

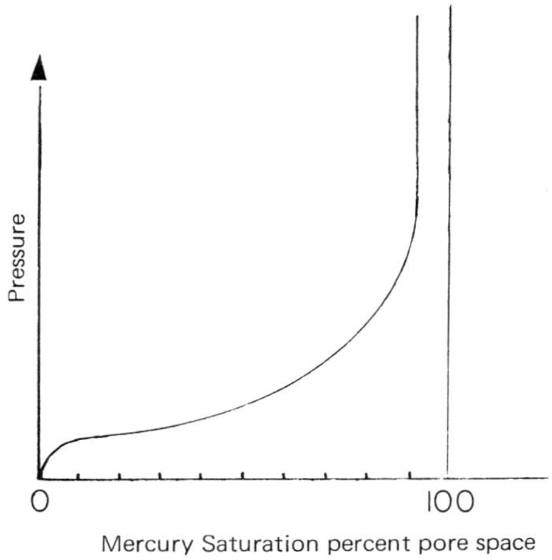

Fig. 14.1. Mercury injection in a medium with matrix porosity.

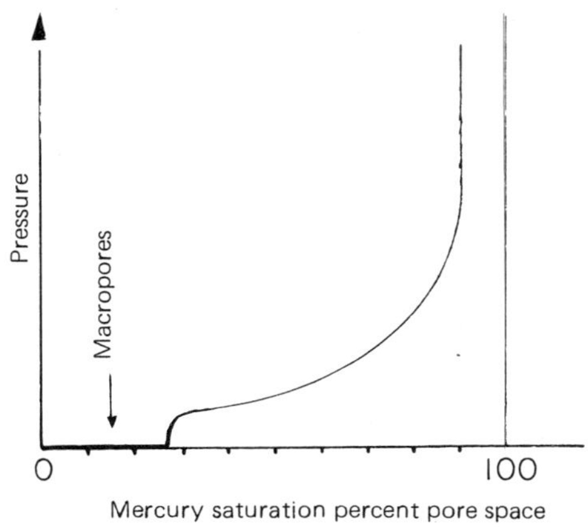

Fig. 14.2. Mercury injection in a medium with matrix porosity and macropores.

15. CASE OF UNCONSOLIDATED MEDIA

The different methods examined above concern consolidated samples. In some cases it is necessary to measure the porosity of relatively well consolidated sands and sandstones which can reach 40% or even more. It is very difficult to obtain a total volume with accuracy since the geometrical shape is practically impossible to define, and when measurements are made with a mercury pump or "voltomètre" there is considerable mercury invasion. In addition, it may be impossible to measure solid volume by immersion because of cleavages within the sample. For unconsolidated sands or sandstones, the best measurement of porosity is, again, by means of the fluid summation method provided that the sample is fresh or preserved.

In the case of marls which are impermeable but have very high porosities, there are several difficulties in the measurement of the V_S by immersion since the marls have a tendancy to crack in the solvent. It is preferable to utilize the gas expansion method (paragr. 13.242).

Remark: In general marls are not analysed.

16. DETERMINATION OF POROSITY BY INDIRECT METHODS (A POSTERIORI ON THE BASIS OF LOGS: ELECTRIC-NEUTRON-GAMMA/GAMMA LOGS)

The companies which make these logs (*Société de Prospection Electrique Schlumberger, PGAC, Lane Wells, Guyod,* etc.) supply collections of charts which make it easy to determine ϕ on the basis of certain types of logs.

16.1. Formation Factor: FF

The matrix of a rock which does not contain clay is an insulator. The electrical conductivity of this rock is due solely to the conducting network formed by the interstitial water contained in the pores (which intercommunicate) and, more exactly, by the shape of this network. For a given rock sample, there is a constant ratio between the resistivity R_o of rock 100% saturated with conducting brine and the resistivity R_w of this brine. This constant which was first introduced by Archie is called the Formation Factor, *FF*.

We have:

$$FF = \frac{R_o}{R_w} \qquad \text{(Eq. 16.11)}$$

FF depends upon the lithological characteristics of the rock and the effective porosity.

In reality, *FF* is not exactly constant when brine is utilized. It should be noted that the formation factor generally increases with salinity. As a result of the fact that the ratio $\frac{R_o}{R_w} \neq$ Const., it becomes necessary to introduce the concept of apparent formation factor $(FF)_a$.

We can write:

$$\frac{1}{R_o} = \frac{1}{R_z} + \frac{1}{(FF)\,R_w}$$

where (FF) is the real formation factor and R_z is rock resistivity.

Thus

$$\frac{1}{(FF)_a} - \frac{1}{(FF)} = \frac{R_w}{R_z}$$

Two measurements with different brines are necessary for obtaining: $\frac{1}{(FF)}$ and $\frac{1}{R_z}$.

Archie has shown that (FF) and ϕ are connected by an equation of the form:

$$(FF) = \frac{a}{\phi^m} \qquad \text{(Eq. 16.12)}$$

where a and m are constants characterizing the rock.

In general, a is close to 1 and $1.3 < m < 2.2$ and 2.6 for compact limestones (Fig. 16.11).

Humble Co. gives the following as a statistical law:

$$FF = \frac{0.62}{\phi^{2.15}} \qquad \text{(Eq. 16.13)}$$

In practice, the following are utilized:

(a) For soft formations ($\phi > 15\%$) : $FF = \frac{0{,}75}{\phi^2}$.

(b) For hard formations ($\phi < 15\%$) : $FF = \frac{1}{\phi^2}$.

Within their range of application these formulae are not very different from the Humble curve. Figure 16.12 shows $FF = f(\phi)$.

Thus since FF is determined on the basis of electric logs, ϕ can be determined on the basis of the above formulae.

16.2. Determination of ϕ by macroresistivity logs

If the formation is 100% saturated with water whose resistivity R_w is calculated on the basis of the SP curve or measured directly, and the resistivity R_o of this formation is estimated on the basis of the resistivity curve, it is possible to deduce FF and ϕ.

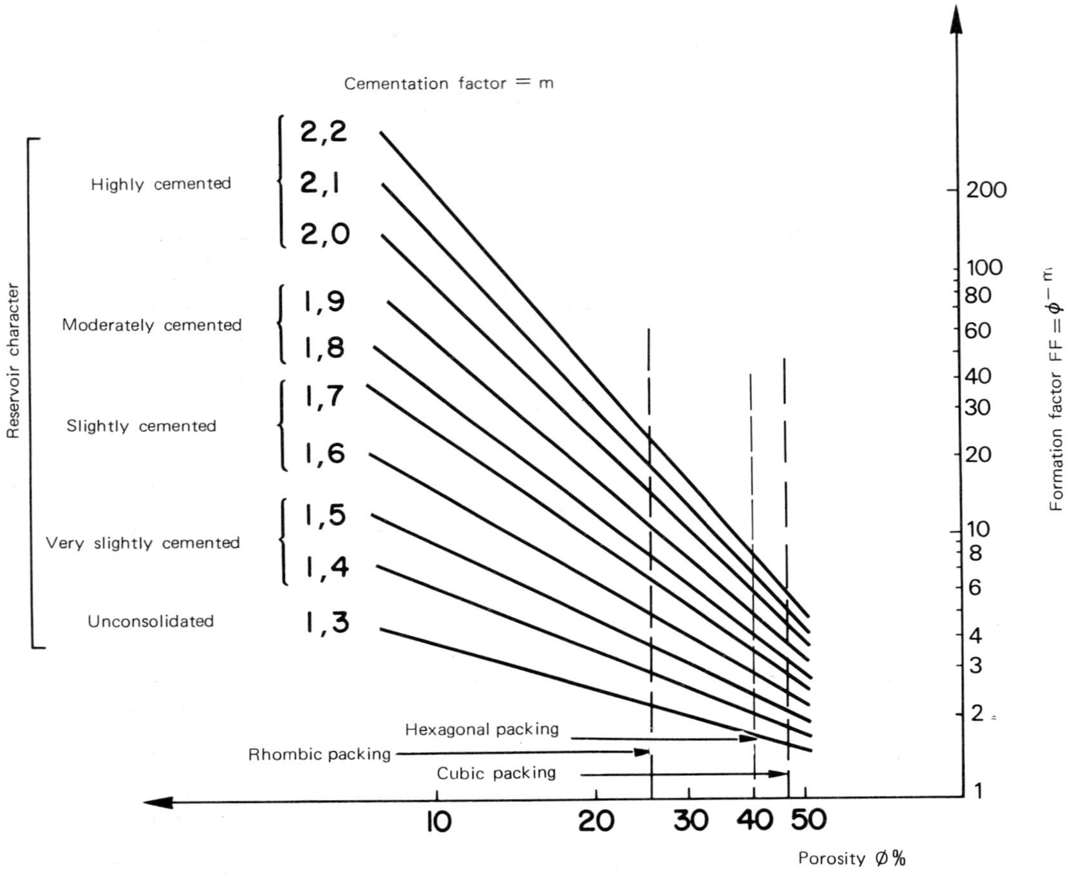

Fig. 16.11. Formation factor versus per cent porosity, for various reservoir characters or cementation classes (from Pirson).

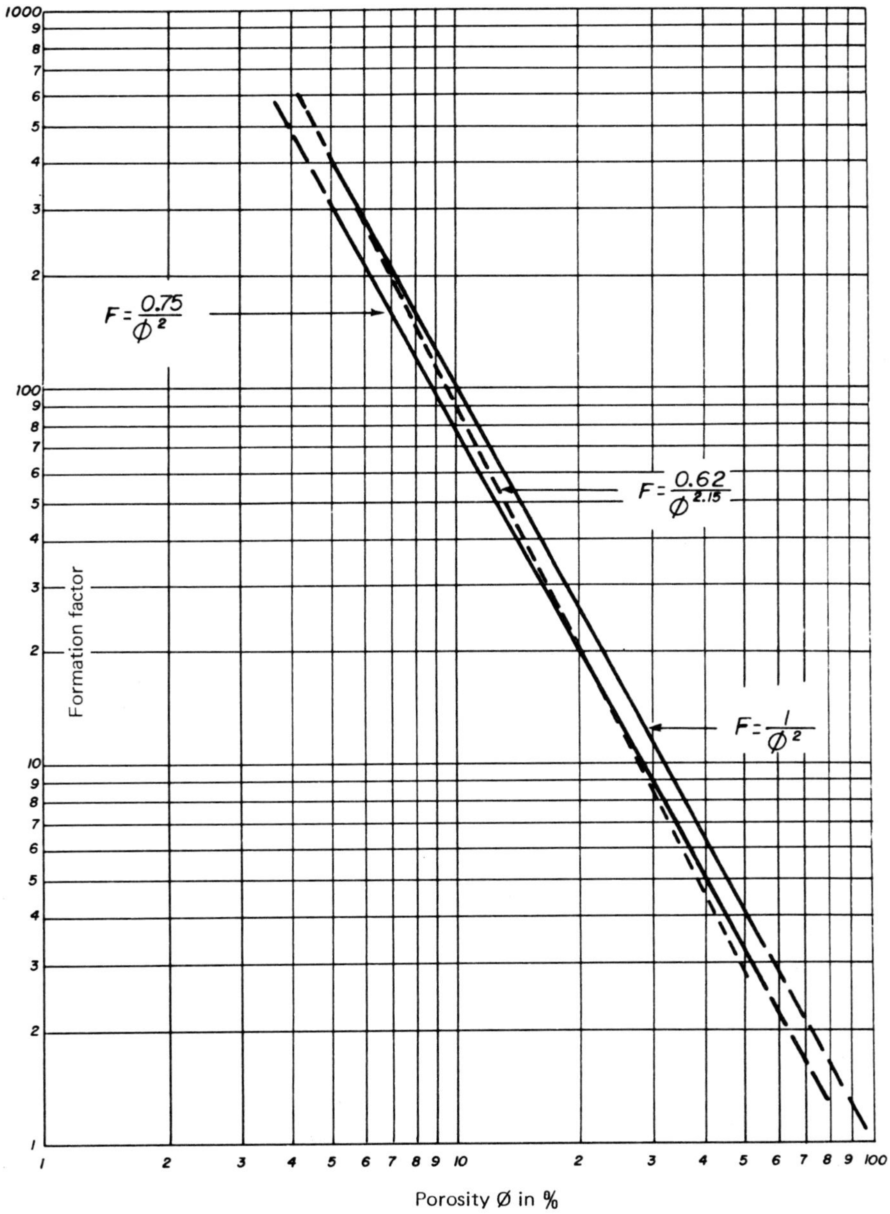

Fig. 16.12. Formation factor. Porosity relationship.

It the formation contains oil, its resistivity increases. Tixier introduces the resistivity R_i of the zone which is invaded by mud filtrate and the true resistivity of the uncontaminated formation R_t.

16.3. Determination of ϕ by microresistivity logs

The following logs are utilized:

(a) Micrologging for soft formations.
(b) Microlaterologging for hard formations.

The resistivity of a part of the formation located near the borehole is measured. It can generally be supposed that the initial brine has been replaced by filtrate.

If the formation is water-bearing, it can be deduced that:

$$(FF) = \frac{R_{xo}}{R_{mf}} \qquad \text{(Eq. 16.31)}$$

where
R_{xo} is the resistivity deduced from the microresistivity curve,
R_{mf} is the resistivity of the filtrate measured on the sample.

If the formation is oil-bearing with a residual oil saturation S_{or} whose order of magnitude is known (through core analysis for example), we have:

$$FF = \frac{R_{xo}}{S_{xo}^2 \cdot R_{mf}} \qquad \text{(Eq. 16.32)}$$

where S_{xo} is the water saturation of the invaded zone: $S_{xo} = 1 - S_{or}$.

Remarks:

(a) The method has the advantage that it is not necessary to collect cores.
(b) The method gives a practically continuous log taking into account rock heterogeneity.
(c) It is not possible to evaluate the error numerically.
(d) In clayey formations, the clay participates in conductivity and FF no longer has a precise meaning when the rock contains more than 5% clay.

16.4. Determination of porosity by means of neutron logging

In neutron logging the reactions of formations under bombardment by fast neutrons are recorded. It is possible to measure secondary gamma radiation produced by the absorption of neutrons by the atoms and the neutron-gamma log is obtained. It is also possible to detect only the slow neutrons resulting from the slowing down of the fast neutrons by hydrogen atoms, and this gives the neutron-neutron log.

In each case the result is a decreasing function of the quantity of hydrogen atoms contained in the formation.

On the supposition that the formation is clean (without clay) and contains water and oil, the neutron curve can in this case be considered a porosity curve.

In reality, experience and theory indicate an exponential relation between the results and the quantity of hydrogen contained in the rock (Fig. 16.41). As a first approximation we have:

$$\log \phi_N = AN + B$$

N = API neutron deflection,
A, B = constants.

The response is affected by various factors, i.e. lithology, borehole diameter.

Sampling is necessary. It should not be forgotten that sandstones can have porosities of the order of 1 to 2%, while porosities can exceed 40% for marls.

The method is applied *in situ* in wells which are either cased or not, and for oil as well as water-bearing formations with the exception of gas layers.

As is the case for all measurements based upon logs it is impossible to give figures for the errors. If the formation contains clay, the data supplied are strictly qualitative.

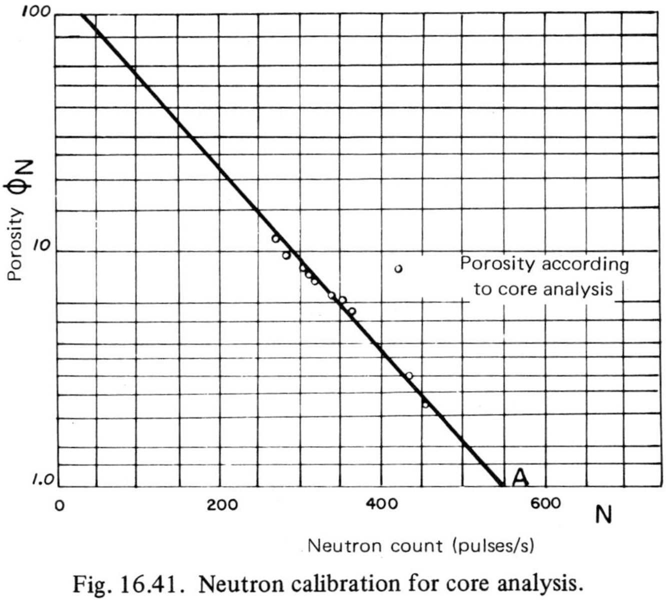

Fig. 16.41. Neutron calibration for core analysis.

16.5. Determination of porosity by means of gamma-gamma logging

Gamma-gamma logging is, firstly, a measurement of formation density. The response of the instrument (density logger) is inversely proportional to the density of the soil traversed by the radiation.

For a given thickness of a certain soil, variations in density indicate variations in porosity.

It is possible to estimate porosity by means of the formula:

$$\phi = \frac{D_g - D}{D_g - D_\phi} \qquad \text{(Eq. 16.51)}$$

where

D = formation density according to the gamma-gamma log,
D_g = matrix density (grain density),
D = density of fluid in pores.

16.6. Determination of porosity by acoustic logging

The velocity of propagation of sound in rocks is affected by the porosity. The velocity of sound falls 60% in a reservoir rock when porosity varies from 3 to 30%.

The following empirical equation has been proposed:

$$\frac{1}{V_{\log}} = \frac{\phi}{V_f} + \frac{1 - \phi}{V_m} \qquad \text{(Eq. 16.61)}$$

where

ϕ = porosity,
$V_{\log}$ = velocity in the rock plotted on the log,
V_f = velocity in the fluid contained in the porosity,
V_m = velocity in the matrix.

The velocity of sound in the rocks (transit time) can easily be measured in the laboratory, in general on vertical samples. This velocity of sound depends upon:

(a) The nature of the rock.
(b) Rock porosity.
(c) Grain density, size, distribution and orientation.
(d) Pore size and distribution.
(e) Grain cementation and stratification.
(f) The nature of the fluids saturating the rock: density, viscosity.

(g) The elastic properties of the rock.
(h) Temperature and pressure.
(i) The measurement instrument (when functioning at low frequencies).

The curve in Fig. 16.61 shows a relation between the velocity of sound and rock porosity for a given type of rock.

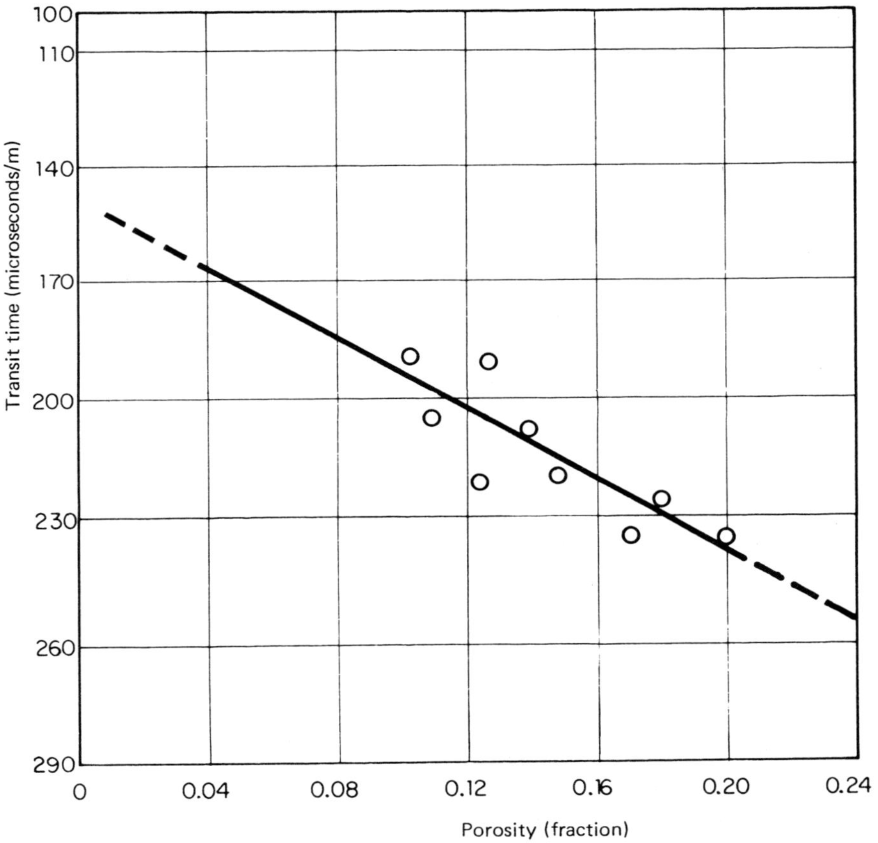

Fig. 16.61. Transit time (measured in the laboratory) versus porosity (for a given rock type).

The table below shows some values for the velocity of sound in three rectangular directions for different rocks and some velocities of sound in several fluids.

VELOCITY OF SOUND (m/s) IN DIFFERENT DIRECTIONS
(AMBIENT TEMPERATURE AND ATMOSPHERIC PRESSURE
ACCORDING TO CORE LABORATORIES)

Rock	Velocity of sound (m/s) according to direction		
	X	Y	Z
Anhydrite (SO_4Ca)	6 200	6 340	6 210
Calcite (CO_3Ca)	7 030	6 570	4 800
Dolomite (CO_3Ca, CO_3Mg)	7 610	–	–
Gypsum (SO_4Ca, 2 H_2O)	5 780	5 320	6 490
Salt	4 680	–	–
Quartz	5 380	5 410	6 520

17. EFFECT OF COMPACTION ON POROSITY

The porosity of sedimentary rocks is a function of the degree of their compaction. The forces of compaction are a function of the maximum depth at which the rock is found. Figures 17.1a and 17.1b show the effect of depth or compaction on porosity (according to Krumbein and Sloss).

Deeply buried sediments have a lower porosity than sediments located at lesser depths. Geertsma considers three kinds of compressibility:

(a) Compressibility of the matrix rock: changes in solid volume in terms of pressure changes.

(b) Compressibility of the total volume: changes in the total volume in terms of pressure changes.

(c) Compressibility of porous volume: changes in pore volume in terms of pressure changes.

Decompression of the field leads to changes in internal rock strains. This change in constraints leads to changes in grain volume in the solid volume and in pore volume.

These changes in volume lead to slight reductions in rock porosity of the order of 0.5% for a variation in pressure of 70 bars in internal fluid pressure, i.e. porosity falls from 20% to 19.9%.

Studies by Van der Knaap skow that this change in porosity for a given rock depends only on the difference between internal and external pressure and not on the absolute value of the pressures.

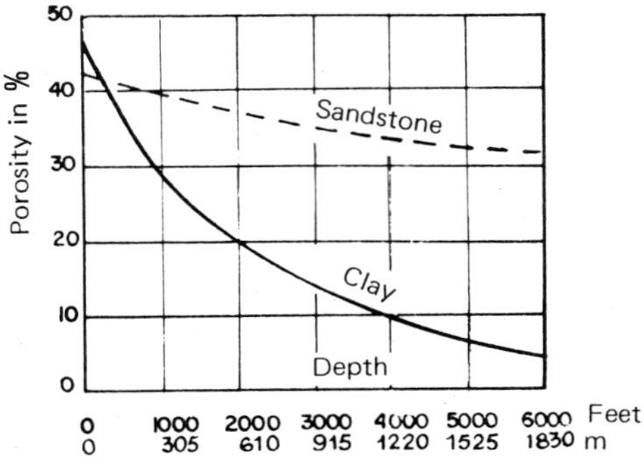

Fig. 17.1a. Effect of natural compaction on porosity (from Krumbein and Sloss).

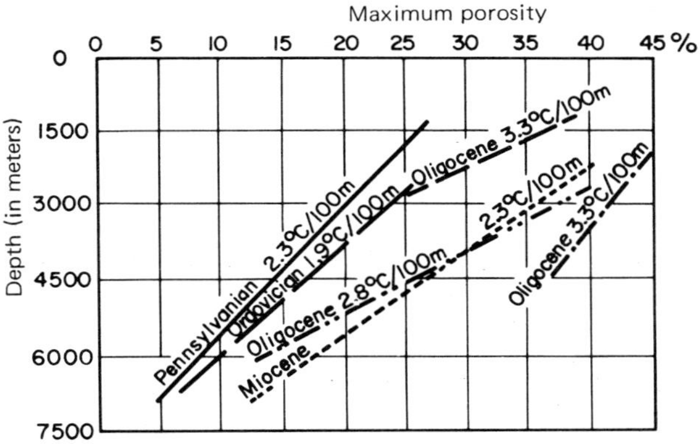

Fig. 17.1b. Graph showing variations in maximum sandstone porosity versus depth, geothermal gradient and age (from J.C. Maxwell).

The compressibility factor $\frac{1}{V_T} \cdot \frac{dV_T}{dp}$ for the limestones and sandstones varies from $0.29 \cdot 10^{-4}$ to $3.63 \cdot 10^{-4}$ bar^{-1}. The compressibility factor expressed in pore volume variation per unit of pore volume and per unit of pressure is equal to the preceding factor divided by the porosity expressed as a fraction.

Hall has shown that the compressibility factor for pore volume is a function of porosity. Figure 17.2 concerns a group of sandstone and limestone samples under the following conditions:

(a) Constant external pressure: 207 bars rel.

(b) Internal pressure varying from 0 to 103.5 bars rel.

The average rock compressibility factors are plotted in terms of porosity.

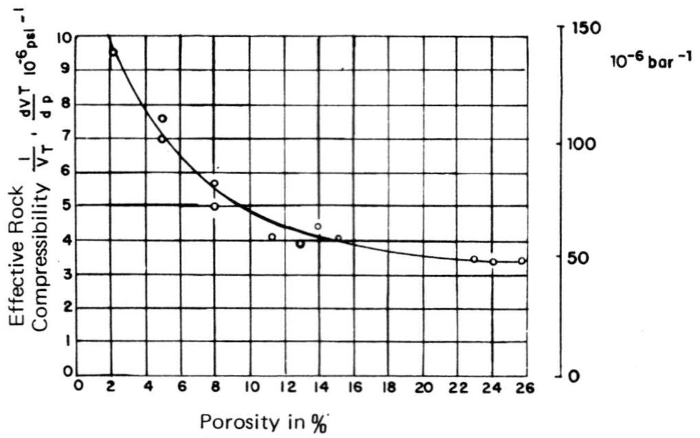

Fig. 17.2. Effective formation (rock) compressibility (from Hall).

Fatt has made measurements of pore volume compressibility for eight sandstones (Fig. 17.3) in terms of the difference in pressure between the external pressure and 85% of the internal pressure. He notes that pore compressibility is a function of the pressure, but for a given scale of porosities studies (10 to 15%) he was not able to establish a correlation with porosity.

Although the compressibility factor for rocks is low, its effects can be of importance in certain reservoir calculations where fluids have compressibility factors of the order of $0.5 \cdot 10^{-4}$ to $3.6 \cdot 10^{-4}$ bar^{-1}.

For example, for the field in Weber sandstones we have:

Depth = 1 524 m.
Geostatic pressure = 0.226 bar/m → 345 bars rel.
Initial field pressure = 176.6 bars rel.

So that:

$$P_{\text{geostatic}} - 0.85\, P_{\text{internal}} = 345 - 0.85 \cdot 176.6 = 193.7 \text{ bars}$$

According to Fig. 17.3 curve *E*, we have:
Rock compressibility factor at 193.7 bars $= 1.23 \cdot 10^{-4}$ (bar)$^{-1}$.

If the field pressure falls to 82.8 bars, the compressibility factor becomes 0.96 10^{-4} (bar)$^{-1}$ so that there is a relatively large variation.

In addition, Geertsma indicates that if the field is not subjected to a uniform external pressure as laboratory samples are, the effective value in the field of the compressibility factor is below the measured value.

The effects of compaction are felt to a considerable extent in measurements of the formation factor and of the resistivity ratio. In general, these measurements are made at atmospheric pressure and laboratory temperature. Experience shows that the formation factor increases with the geostatic pressure (Fig. 17.4), and the formation factor/porosity plot is shown in Fig. 17.5.

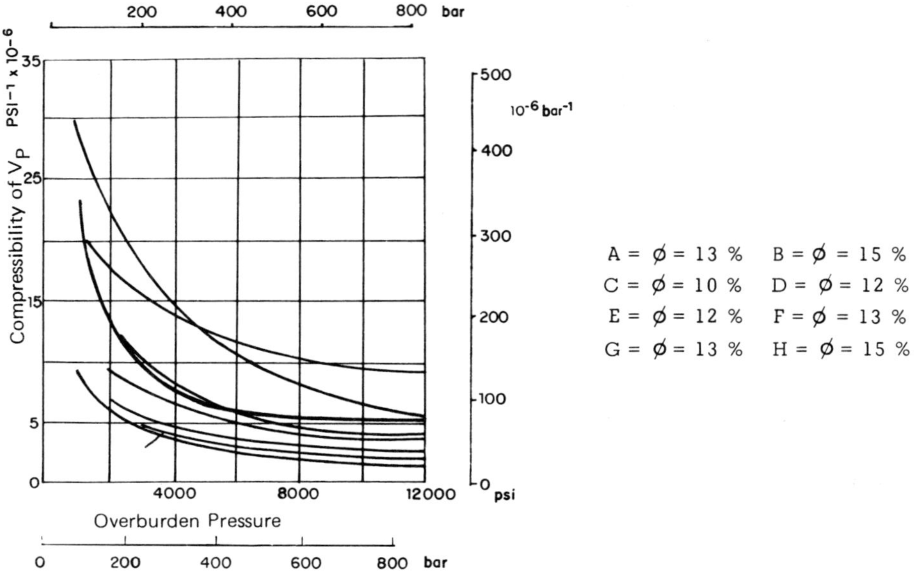

Fig. 17.3. Pore volume compressibility factor for 8 sandstones in terms of overburden pressure (from Fatt).

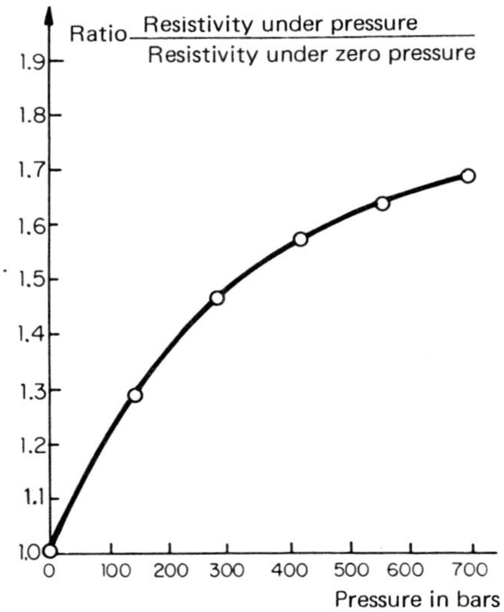

Fig. 17.4. Effect of pressure on resistivity.

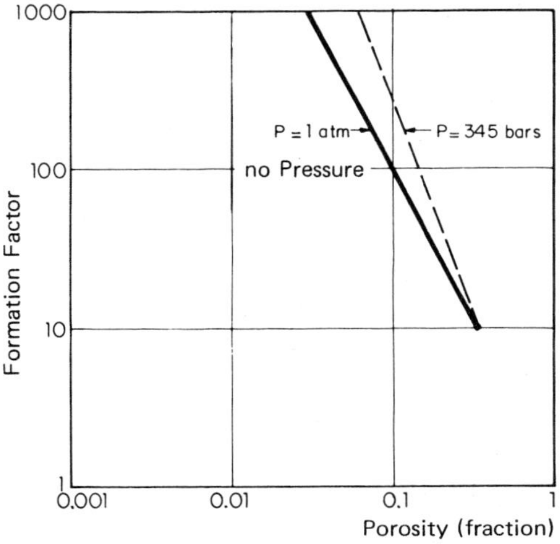

Fig. 17.5. Effect of overburden pressure on formation factor.

18. COMPARISON OF THE VARIOUS METHODS UTILIZED FOR POROSITY MEASUREMENTS

The methods used vary depending upon:

(a) The nature of the samples (fresh or extracted).
(b) The degree of consolidation of the formation and its nature.
(c) The size of samples.

. Conventional analysis of small samples-plugs- (10 cm^3): **Conventional Core Analysis.**
. Conventional analysis of whole cores (in general fresh): **Whole Core Analysis.**
. Conventional analysis of small samples collected by side wall coring: **Side Wall Core Analysis.**
. Special analysis of cores (V_T = 50 to 80 cm^3): **Special Core Analysis.**

These different methods of core analysis will be studied in detail in Chapter 2.

The table below gives the advantages and disadvantages of each of the most frequently employed methods.

19. APPENDIXES I AND II

Appendix I. Mechanical properties of reservoir rocks

The study of the mechanical properties of rocks has been the subject of many publications. In this course will be found only the usual values for the mechanical characteristics of sedimentary rocks (Table 19.1).

It is known that:

1. The modulus of static elasticity E is defined by the slope of the stress-strain curve.

2. The coefficient of Poisson ν is by definition equal (except for the sign) to the ratio of horizontal strains ϵ_3 (radial strain) and vertical strains ϵ_1 (axial strain):

$$\nu = -\frac{\epsilon_3}{\epsilon_1}$$

3. Simple compression strength R_c is the unconfined breaking strength.

4. Tensile strength which is highly influenced by rock cracks is generally determined:

(a) Either by tensile tests (not very practical).
(b) Or by diametrical compression strength tests (Brazilian test).

TABLE OF COMPARISONS OF VARIOUS METHODS
FOR ESTABLISHING POROSITY

Methods	Advantages	Limitations
Summation of fluids	Utilizes large, fresh samples Rapid Exact (relatively exact $\phi\,\% \pm 0.5$).	Water and oil are determined after gauging Presence of dehydratable minerals The samples analysed cannot be utilized afterwards
Mariotte-Boyle law (measurement of V_P)	Quite rapid in most cases The samples can be used for other measurements Exact (reproductibility 2% measured ϕ)	Slow for low permeabilities The drying technique is of importance particularly when dehydratable minerals are present Measurement of the total crystallisable volume as in the other methods
Measurement of air in pores (Vellinger or Washburn-Bunting porosimeter)	Accuracy is high if the various technical operations have been well performed (ϕ reproducible to $\pm$ 1 point for $8\% < \phi < 40\%$).	Technique to be utilized with care Instruments must be very clean The sample cannot be used for other measurements The sample must be carefully prepared (as for the preceding technique)
Resaturation (weighing of the liquid filling pores)	Determination of porosity during the preparation of samples for other measu-rements	Accuracy limited because sa-turation is more or less complete Same remarks for the prepa-ration of samples as for the second case above
Grain density	Determinaion of total porosity	Slow and difficult method Laboratory technique

TABLE 19.1

MECHANICAL CHARACTERISTICS OF SOME SEDIMENTARY ROCKS
(according to Mr. Le Tirant *et al.* : Manual of hydraulics fracturation – ARTEP Editions Technip Paris, 1972)

Nature and origin of rock	Young nodulus E (bar)	Poisson's ratio ν	Unconfined compressive strength R_c (bar)	Tensile strength R_T (bar)	Sound velocity (m/s)	Compressibility C_b (10^{-6} 1/bar)
Hassi-Messaoud sandstone	300 000 to 500 000	0.14 to 0.21	1 100 to 1 250	20 to 90		4 to 6
El Agreb sandstone	400 000 to 550 000	$\simeq$ 0.2	1 350 to 1 550			4
Zarzaïtine sandstone	450 000					
Fine Vosges sandstone	125 000		$\simeq$ 300			
Coarse Vosges sandstone	235 000		400	30 to 50		
Fontainebleau sandstone	300 000 to 400 000	0.15 to 0.25	600 to 1 900	$\simeq$ 50		4 to 6
Clayey sandstone (35% clay)	50 000 to 90 000		700 to 740			
Bituminous sandstone	30 000 to 60 000	0.25 to 0.30	160 to 260			$\simeq$ 30
Saint-Maximin limestone	66 000 to 82 000	0.19 to 0.25	90 to 120	10 to 12	2 300 to 2 600	20 to 25
Rouffach cornstone			450 to 700			
Marquise sandstone	775 000 to 950 000	0.28 to 0.33	1 100 to 1 500	100 to 140	6 200 to 7 000	1.5
Marl	$\simeq$ 80 000 to 100 000	0.41				6
Tersanne salt	$\simeq$ 50 000	0.36	150 to 200			15 to 20

Appendix II. Effect of compaction on pore volume

Example of the utilization of an effective compressibility factor: rocks + fluids.

For an undersaturated Hassi-Messaoud type oil reservoir:

Field pressure = P_G	400 bars
Saturation pressure = P_B	200 bars
Irreducible water saturation = S_{iw} (S_{cw})	30 % pore volume
Porosity = ϕ	10 %
Water compressibility factor = c_w (water)	$4.5 \cdot 10^{-5}$ bar^{-1}

Initial volumetric factor: B_{oi} — 1.760 vol/vol at 400 bars
Volumetric factor at 300 bars: B_o — 1.810 vol/vol at 300 bars

We calculate:

At 400 bars:

(a) pore volume — 0.10 m^3/m^2 · m of rock volume
(b) volume of connate water — 0.03 m^3/m^2 · m of rock volume
(c) volume of oil in place — 0.07 m^3/m^2 · m of rock volume

At 300 bars:

(a) pore volume — $0.10\,(1 - c_p \cdot \Delta p)$

According to Hall, $c_p = 6.68\ 10^{-5}$ bar^{-1} = compressibility factor

so that pore volume — $0.10\,(1 - 6.68 \cdot 10^{-5} \cdot 100) =$ 0.09933 m^3/m^2 . m

(b) connate water — $0.03\,(1 + c_w \cdot \Delta p) =$ 0.030135 m^3/m^2 . m

(c) oil volume — $0.09933 - 0.03013 =$ 0.06920 m^3/m^2 . m

So that the oil in place:

(a) At 400 bars: $\dfrac{0.07}{1.760} = 0.03977$ m^3/m^2 · m

(b) At 300 bars: $\dfrac{0.0692}{1.810} = 0.03823$ m^3/m^2 · m

Difference (400 – 300) = 0.00154 m^3/m^2 · m.

Oil recovery:

$$\frac{0.00154}{0.03977} \cdot 100 = 3.87\%$$

Average oil compressibility in the 400/300 kg/cm^2 pressure range is:

$$c_o = -\frac{1}{V}\frac{\Delta V}{\Delta P}$$

$$c_o = \frac{V_o - V_{oi}}{V_{oi}\,(P_i - P)} = \frac{B_o - B_{oi}}{B_{oi} \cdot \Delta P} = \frac{1.810 - 1.760}{1.760 \cdot 100} = 28.4 \cdot 10^{-5}\ \text{bar}^{-1}$$

Recovery: N_p for N in place:

$$\frac{N_p}{N} = c_e \cdot \frac{B_{oi}}{B_o} \cdot \Delta p$$

c_e effective rock and fluid compressibility (composite).

So that:

$$c_e = c_o + c_w \cdot \frac{S_{cw}}{1 - S_{cw}} + c_p \cdot \frac{1}{1 - S_{cw}}$$

$$c_e = 28.4 \cdot 10^{-5} + 4.5 \cdot 10^{-5} \cdot \frac{0.3}{0.7} + 6.68 \cdot 10^{-5} \cdot \frac{1}{0.7} =$$

$$39.87 \cdot 10^{-5}\ \text{bar}^{-1}$$

$$\frac{N_p}{N} = 39.87 \cdot 10^{-5} \cdot \frac{1.760}{1.810} \cdot 100 = 0.0265 = 3.87\%$$

The same calculation can be made leaving out rock and water compressibility. We have:

$$\frac{N_p}{N} = 28.4 \cdot 10^{-5} \cdot \frac{1.760}{1.810} \cdot 100 = 2.76\%$$

This shows the importance of composite c_e above the bubble point.

Any study of oil recovery from undersaturated reservoirs requires that the rock compressibility factor be known. It is determined by core measurements.

The fluid compressibility factor is supplied by PVT studies.

2

permeability of reservoir rocks

21. DEFINITION. THEORY. DARCY'S LAW

21.1. Definition of permeability and Darcy's law

Permeability characterizes the ability of rocks to allow the circulation of fluids contained in their pores.

While studying water filtration, Darcy showed experimentally that the rate of flow per surface area $\frac{Q}{S}$ of a filter was, all other factors being equal, proportional to the head between the two faces ΔH:

$$\frac{Q}{S} = a\Delta H$$

This is Darcy's law in its original form.

The result which it expresses was later further specified by the introduction of fluid viscosity which makes it possible to define the notion of permeability, and it was generalized by the substitution of potentials for head.

For a thin slice with the thickness dx and cross-section S traversed perpendicularly at its faces by the rate of flow Q (counted in volume at the temperature and average pressure of the slice) of a fluid with the viscosity μ affected by a pressure difference dp, Darcy's law in its generalized form is written as follows (not taking into account gravitational force):

$$\mathrm{d}p = \frac{\mu}{k} \cdot \frac{Q}{S} \mathrm{d}x$$

or

$$Q = \frac{Sk}{\mu} \cdot \frac{\mathrm{d}p}{\mathrm{d}x} \qquad \text{(Eq. 21.11)}$$

with the coefficient k being the permeability of the slice (Fig. 21.11).

Darcy's law supposes that:

(a) There is no reaction between fluid and rock.
(b) There is only one fluid present.

Permeability is reduced if there is a reaction between fluid and rock. There is also a reduction in permeability for each phase when several fluids are present.

Permeability depends upon pore dimensions and configuration. Sandstones with large pores have high permeability, while a very fine grained sandstone has very low permeability; an oölitic limestone will have high permeability and an intercrystalline limestone will have very little permeability.

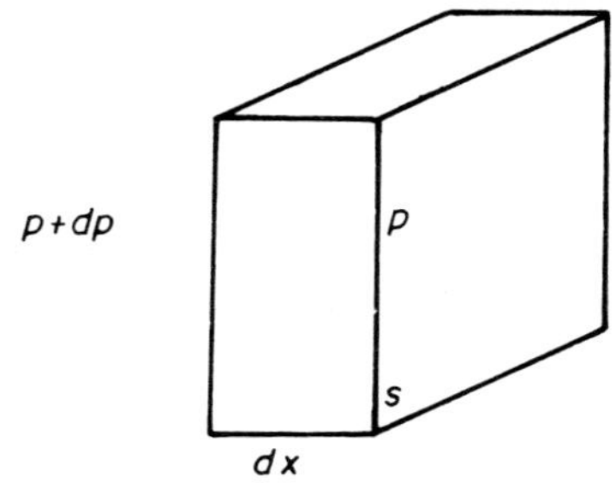

Fig. 21.11.

In a naturally porous medium, permeability can vary with the **direction of flow.** The **statistical nature of permeability** must be taken into account, and this cannot be done for a point since this characteristic must always be considered as linked to a surface element with much larger dimensions than the channel sections.

21.2. Units

The Darcy equation (Eq. 21.11) shows that permeabilities are homogeneous with surface areas.

(a) CGS unit = perm = 1 cm².

If:

$$Q = 1\ \text{cm}^3/\text{s} \qquad \mu = 1\ \text{P}$$

$$S = 1\ \text{cm}^2 \qquad \frac{\Delta P}{\Delta x} = 1\ \text{barye/cm}$$

then $k = 1$ perm.

(b) MKSA unit = m².

The perm is too large a unit for porous media.

The practical unit is the darcy and its sub-multiples (especially the millidarcy = 10^{-3} darcy).

The darcy is the permeability of a medium allowing the passage of 1 cm^3/s of a fluid whose viscosity is 1 cP with a pressure gradient of 1 atmosphere per centimeter through a surface area of 1 cm^2.

Thus:

$$k = \frac{\mu Q}{S} \cdot \frac{dx}{dp} \qquad \text{(Eq. 21.21)}$$

and by definition

$$1 \text{ darcy} = \frac{1 \text{ cP} \cdot 1 \text{ cm}^3/\text{s}}{1 \text{ cm}^2} \cdot \frac{1}{1 \text{ atm/cm}}$$

so that 1 darcy = $0.9869 \cdot 10^{-8}$ perm.

In practice:

$$\begin{aligned} 1 \text{ darcy} &= 10^{-8} \text{ perm} \\ 1 \text{ millidarcy (mD)} &= 10^{-11} \text{ perm} \end{aligned}$$

In hydraulics permeability unit is the centimetre per second and 1 darcy $\simeq$ $0.9615 \cdot 10^{-3}$ cm/s.

Remark:

A very broad range of permeabilities is found varying from 0.1 mD to more than 10 and 15 darcys.

21.3. Slippage at the walls: Klinkenberg effect

Permeability measurements for several fluids (gas, liquids) and different pressure gradients give different results (Fig. 21.31). The small transversal dimensions of the channels in which the fluids circulate within the porous medium should be noted.

According to the kinetic theory of gases, the molecules can be considered as tiny spheres with diameters of approximately one ten thousandth of a micron separated by distances which are approximately ten times their own size at atmospheric pressure. These molecules move at very high speeds (of the order of the speed of sound) and collide in a random manner. The average length of this free movement is inversely proportional to the pressure. The average free movement can be very great for rarefied gases.

At pressures close to atmospheric pressures (even for a capacity with dimensions of the order of 1 micron), most of the collisions will still take place between the molecules and not against the walls because of the large number of molecules present in the unit of volume.

On the other hand, at very low pressures, internal friction due to collisions of the molecules with each other tends to disappear and the flow is clearly affected by the relative proportion of collisions with the walls. Molecular flow then occurs and the coefficient of viscosity is no longer meaningful.

From the above, it follows that the simple application of Darcy's law tends to lead more and more to erroneous results as the average flow pressure falls. Klinkenberg has related apparent permeability k_a measured for gas for an average pressure P_m to the true permeability k_L by an equation with the form (Fig. 21.32):

$$k_a = k_L \left(1 + \frac{b}{P_m}\right) \qquad \text{(Eq. 21.31)}$$

where

b = Const. depending upon the average free movement λ of the molecule at P_m.

$$b = \frac{4c' \lambda P_m}{r}$$

with

$c' \approx 1$,

r = channel radius.

Extrapolation to the abscissa $\frac{1}{P_m} = 0$ gives k_L.

Permeability k_L is called liquid permeability.

It was observed that when the gas is rarefied, what can be called the slippage at the walls effect occurs which was discovered by Maxwell and then by Chapman. Klinkenberg transposed these results for the kinetic theory of gases in porous media, and this phenomenon will be called simply the Klinkenberg effect *(Shell Oil Co.)*.

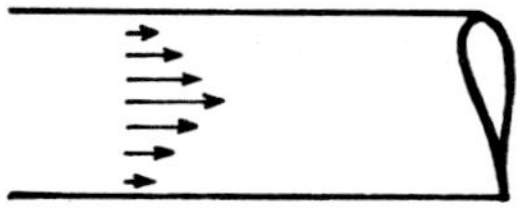

Fig. 21.31. Liquid flow.

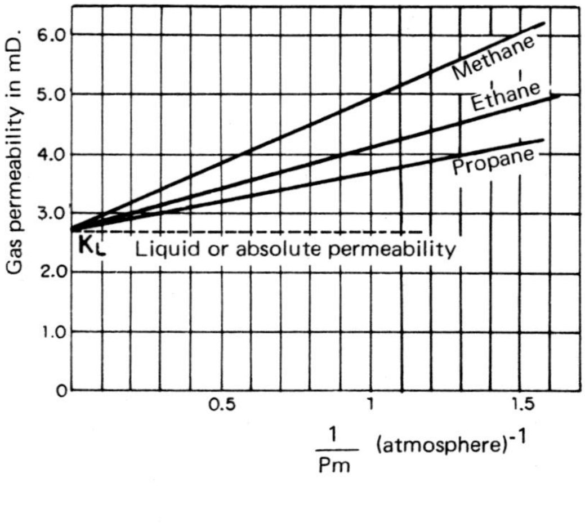

Fig. 21.32. Variation in gas permeability with mean pressure and type of gas (from Klinkenberg).

The average corrections vary approximately from 1% for high k_a values to 70% for very low k_a values. Tables of experimentally determined corrections are used in the laboratory. The value k_L is used for some measurements in special core analysis such as relative permeabilities for water/oil, gas/oil, etc. In many cases the k_a is determined by instruments operating at pressures close to atmospheric pressure. In order to determine the Klinkerberg effect correction and obtain a liquid permeability value, the following can be used:

(a) A formula given by Iffly (pressure P_m being approximately 1 atm):

$$k_L = k_a \frac{c}{c + 0.174}$$

where c in microns is deduced from the average pressure P_m of the invasion plateau of the capillary pressure curve for mercury injection: $c = \frac{7}{P_m}$.

(b) A formula given by Purcell:

$$k_L = \frac{c^2 \phi/P_m^2 + 2k_a - \sqrt{c^2 \phi/P_m^2 \, (c^2 \phi/P_m^2 + 4k_a^2)}}{2}$$

where

k_a = in millidarcys,
ϕ = fraction,
P_m = average measurement pressure in atm,

$$c \begin{cases} = 2.26 \text{ if } 0 < k_a < 10, \\ = 2.42 \text{ if } 10 < k_a < 100, \\ = 2.72 \text{ if } 100 < k_a, \end{cases}$$

(c) A chart used by *Core Laboratories Inc.* (Fig. 21.33).

Remark:

Since the average pressures prevailing in a gas pool are of the order of several hundred bars it is not necessary to correct Darcy's law for this effect even if permeability is low.

21.4. Specific, effective and relative permeabilities (Fig. 21.41)

Specific permeability: this is the permeability measured when there is only a single fluid present, for example air permeability, water permeability or oil permeability.

Effective permeability: when there is a fluid in the rock pores (with saturation other than irreducible minimum saturation) the results of measurement of permeability by means of a second fluid is called the effective permeability for this fluid.

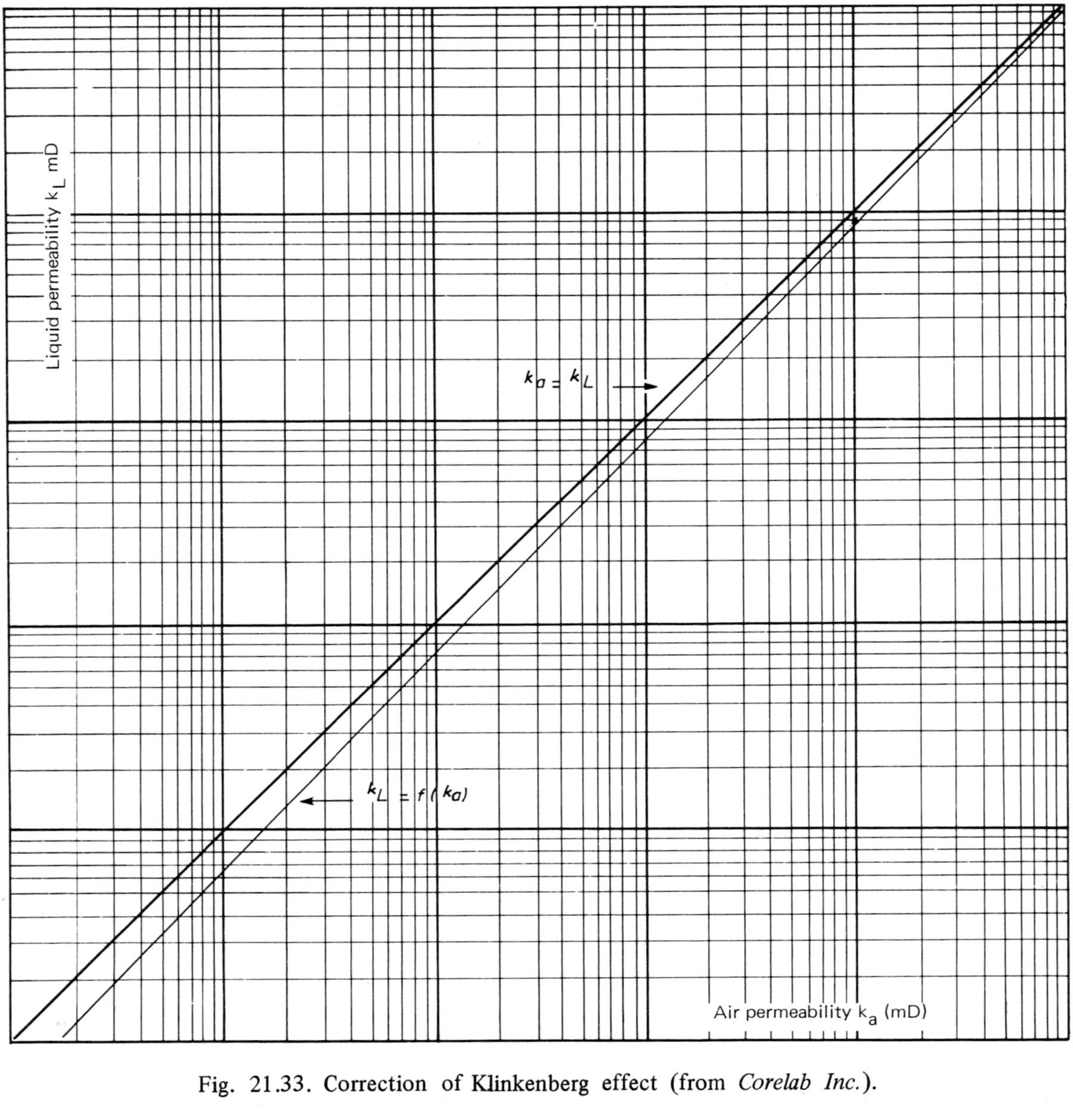

Fig. 21.33. Correction of Klinkenberg effect (from *Corelab Inc.*).

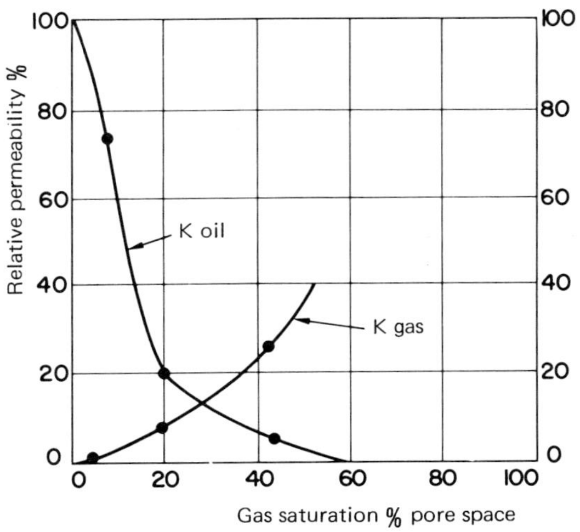

Fig. 21.41. Oil displacement by gas.

Gas saturation = 45 % of pore space
Specific permeability = 250 mD
Effective permeability to oil = 6.2 mD
Effective permeability to gas = 70 mD
Relative permeability to oil = 6.2/250 = 0.025
Relative permeability to gas = 70/250 = 0.28

Relative permeability:

$$\frac{\text{effective permeability}}{\text{specific permeability}}$$

In the case of an oil-bearing formation there are often two fluids present, i.e. gas and oil, or even three, i.e. gas, oil and water.

TYPICAL SPECIFIC AND EFFECTIVE PERMEABILITY DATA

	Clean sand	Sand with clay	Limestone
$k_{air} = k_a$	400	80	1.0
$k_{air\ (Klinkenberg)} = k_L$	370	70	0.6
$k_{water} = k_w$	370	50	0.6
$k_{oil\ with\ connate\ water} = k'_o$	370	50	0.4
$k_{water\ with\ residual\ oil} = k'_w$	70	10	0.05

22. VALIDITY OF DARCY'S LAW

Darcy's law which was established empirically is simply an equation for pressure drop:

$$dp = \frac{\mu Q}{S_k} \cdot dx$$

It should be considered as a particular expression of the principle of the proportionality of the forces of acceleration in the case of porous media as Poiseuille's equation is for capillary tubes. According to conventional fluid mechanics studies, losses of head are not always proportional to the rate of flow.

Darcy's equation defining permeability is linked to laminar flow in porous media. This laminar flow is not always achieved especially in gas flows.

22.1. Real gas flows

Let us suppose that there is a metal tube filled with sand (Fig. 22.11) in which perfect permanent flow exists. Upstream and downstream pressures, respectively P_G and P_F are constant. The mass flow Q_m is the same in all slices and for each slice is related to the volume flow rate Q by:

$$Q_m = Q\rho$$

where

ρ = gas density in the slice where the pressure is p.

We have: $\rho = bp$, with b = Const., where $Q_m = Qbp$

$$Q = \frac{1}{bp} \cdot Q_m$$

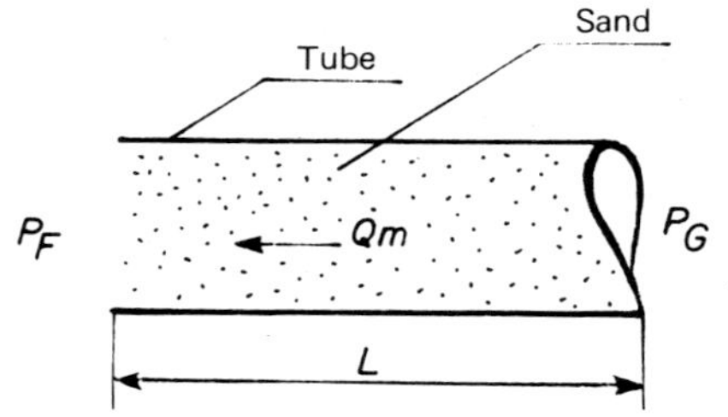

Fig. 22.11. Linear flow of a perfect gas in sand-filled tube.

Taking into account (Eq. 21.11), $Q = \frac{Sk}{\mu} \cdot \frac{dp}{dx}$, we obtain:

$$bp \cdot dp = \frac{\mu Q_m}{Sk} \cdot dx \qquad \text{(Eq. 22.11)}$$

a differential relating p to distance x.

After integration (for thickness L) we have:

$$\frac{b}{2} \cdot (P_G^2 - P_F^2) = \frac{\mu Q_m}{Sk} \cdot L$$

where when measuring Q at pressure p and at flow temperature and taking into account $Q_m = Qbp$

$$Q = \frac{Sk}{\mu L} \cdot \frac{P_G^2 - P_F^2}{2p} \qquad \text{(Eq. 22.12)}$$

The results of this calculation are not in accord with reality unless special precautions are taken.(Eq. 22.12) establishes that Q_m and $(P_G^2 - P_F^2)$ are proportional. Gas sondes show another result which can be explained by introducing irregularities in channel shape (effect on S) and soil heterogeneities (effect on k).

All known gas sondes give an equation with the following form between Q and P:

$$Q = \alpha (P_G^2 - P_F^2)^n \qquad \text{(Eq. 22.13)}$$

where α and n are two constants for a given well (n is between 0.5 and 1).

The origin of (Eq. 22.13) is to be sought for in mechanical flow conditions in porous media rather than in thermodynamic considerations (the flows are practically isothermal and the differences for the real gases in relation to perfect gases cannot justify the usual values for the exponent n).

22.2. Liquid flows

In general, liquids follow Darcy's law (single-phase fluids). The Darcy equation integrated for a fluid with invariable density in circular radial flow (Fig. 22.21) within a layer of constant thickness h and between two boundaries a and R with the respective pressures P_F and P_G leads to the equation:

$$Q = \frac{2\pi hk}{\mu} \cdot \frac{P_G - P_F}{\ln \frac{R}{a}} \qquad \text{(Eq. 22.21)}$$

i.e.

$$Q = I (P_G - P_F) \text{ (Fig. 22.22)} \qquad \text{(Eq. 22.22)}$$

The constant I is the productivity index of the borehole (in relation to the FVF).

In certain cases, the relation $Q = Q\,(P_G - P_F)$ is not linear (Fig. 22.23) but parabolic. This quite rare phenomenon appears in fissured media.

22.3. Attempt to establish a rate of flow law which is valid for both gas and liquids

If a cylindrical porous medium (Fig. 22.31) is considered as a bundle of capillary tubes of the same length, the following similarity can be established:

Darcy's law gives:

$$Q = Sk \cdot \frac{\Delta P}{\mu l}$$

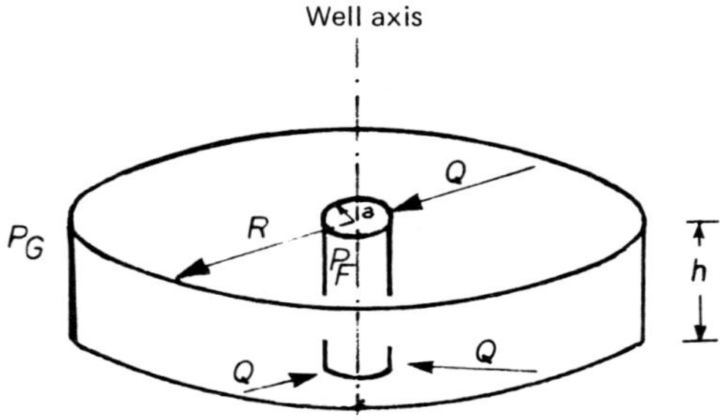

Fig. 22.21. Radial flow of incompressible fluids to central well bore.

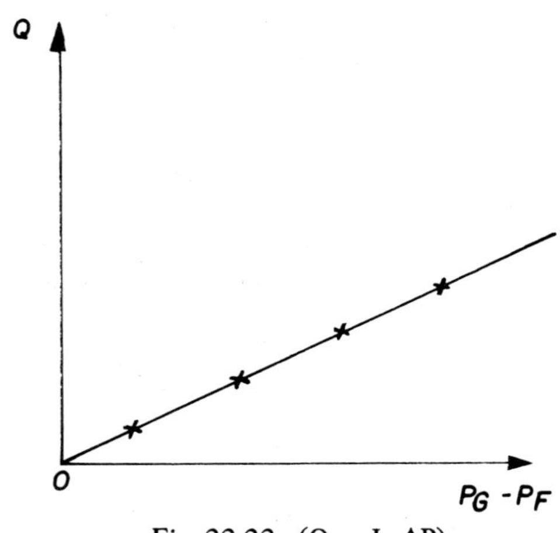

Fig. 22.22. (Q = I . ΔP)

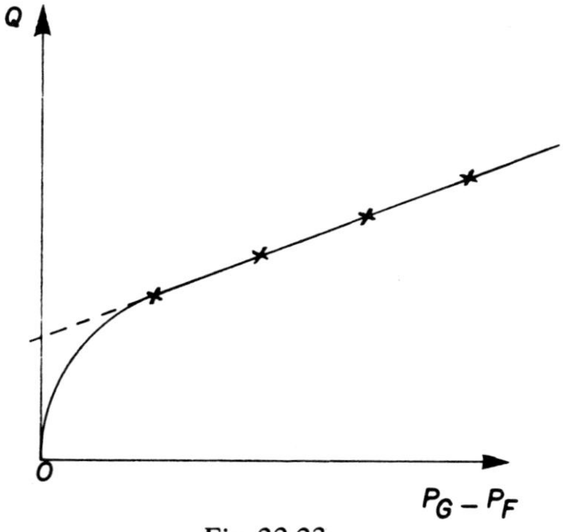

Fig. 22.23.

Poiseuille's low for a capillary tube with the radius r and the same length gives:

$$q = \frac{\pi r^4}{8} \cdot \frac{\Delta P}{\mu l}$$

By correctly choosing the number of capillary tubes n and their radius r it is possible to obtain the same flow rate and pore volume. It is sufficient that:

$$\frac{n\pi r^4}{8} = Sk$$

i.e.

$$r = \sqrt{\frac{8k}{\phi}} \qquad \text{(Eq. 22.31)}$$

for $S = \dfrac{n\pi r^2}{\phi}$, ϕ being porosity.

This analogy gives an idea of pore size and will be useful for taking into account "capillary pressure – porous media saturation" curves (see Chapter 3). However experience has shown that this simplified scheme for porous media is insufficient.

If we examine the sections of a mass made up of glass beads, the basic property of porous media can be observed, i.e. the variations in channel section from one plane to the next. Displacement conditions for a molecule in a channel of this kind will be very different from displacement conditions for a molecule in a capillary tube.

The equation for pressure drop in a porous channel has been studied by A. Houpeurt (1).

In very schematic form, it can be supposed that there are narrow openings along the path of the molecule and that the vein section rises from $2c$ at the throats to $2mc$ at the widenings, with the throats being separated by a distance of $2n''c$. The vein would then follow the path of a sequence of cone frustums (Fig. 22.32).

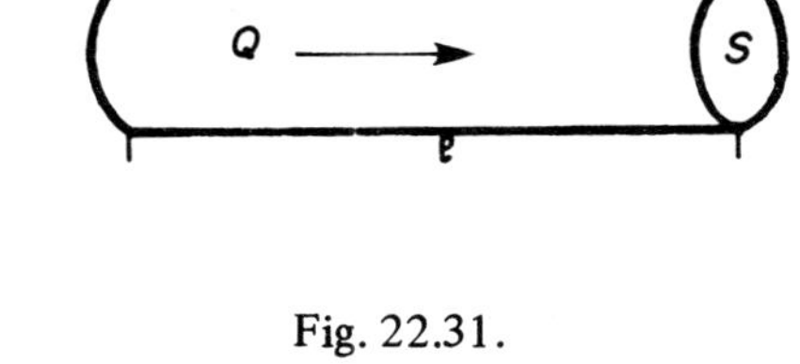

Fig. 22.31.

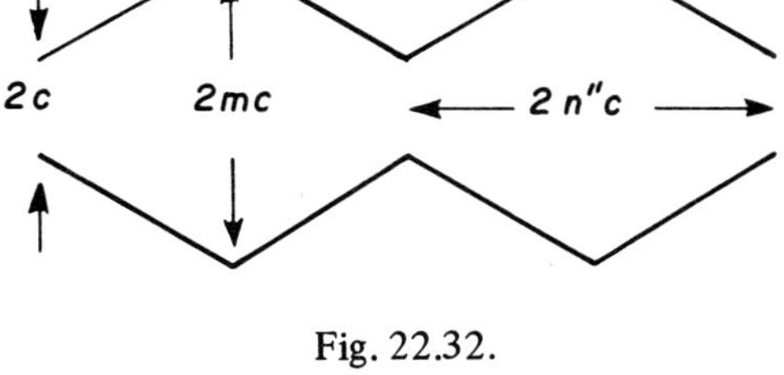

Fig. 22.32.

(1) Houpeurt A., "Sur l'écoulement des gaz dans les milieux poreux". *Rev. Inst. Franç. du Pétrole,* Vol. XIV, Nos 11 and 12, Nov., Dec. 1959.

A simple calculation ([1]) shows that for an elementary channel the pressure drop due to viscous friction between two successive widenings is:

$$\delta_1 p = \frac{8\mu p}{3\pi c^4} \cdot \frac{2n''c}{m^3} (m^2 + m + 1)$$

The pressure drop due to the supposed total dissipation of kinetic energy along the same path is equal to:

$$\delta_2 p = \frac{\rho q^2}{\pi^2 c^4} \left(1 - \frac{1}{m^4}\right)$$

so that the total pressure drop between the two throats is:

$$\delta p = \frac{8\mu q}{3\pi c^4} \frac{2n''c}{m^3} (m^2 + m + 1) + \frac{\rho q^2}{\pi^2 c^4} \left(1 - \frac{1}{m^4}\right)$$

For a distance dr of an order much higher than the dimension c, the pressure drop will be:

$$dp = \delta p \frac{dr}{2n''c}$$

$$dp = \frac{dr}{2n''c} \left[\frac{8\mu q}{3\pi c^4} \frac{2n''c}{m^3} (m^2 + m + 1) + \frac{\rho q^2}{\pi^2 c^4} \left(1 - \frac{1}{m^4}\right) \right]$$

Here we introduce the mass flow rate $qm = \rho q$ for each channel and that for a slice of section S would be $Q_m = N_{qm}$, with N being the number of channels. In order to take into account the empty space volume of the slice it is necessary that:

$$\frac{N}{S} = \frac{3\phi}{\pi c^2 (m^2 + m + 1)}$$

The permeability of the medium can then be calculated by identifying the equation for Darcy's law:

$$dp = \frac{\mu Q_m}{Sk\rho} \cdot dr$$

and the pressure drop due to viscous friction for the same path calculated on the supposition that the velocity along surfaces in turbulent zones is zero, i.e.

$$d_1 P = \frac{8\mu}{3\pi c^4 \rho} \cdot \frac{Q_m}{N} \cdot \frac{1}{m^3} (m^2 + m + 1)\, dr$$

([1]) Houpeurt A., "Etude analogique de l'écoulement radial circulaire transitoire des gaz dans les milieux poreux". *Rev. Inst. Franç. du Pétrole,* Nos 4, 5, 6, April, May, June 1953.

Note : The channel is given the profile of an algebric curve for calculation of pressure drop.

If N is replaced by its value we have:

$$d_1 P = \frac{8\,\mu}{3\pi c^4 \rho} \cdot \frac{Q_m\, \pi c^2\, (m^2 + m + 1)^2}{3\,\phi\, S\, m^3} \cdot dr$$

$$= \frac{\mu Q_m\, dr}{S\rho} \cdot \frac{8\,(m^2 + m + 1)^2}{9\,\phi\, c^2 m^3}$$

and consequently by identification with Darcy's law

$$k = \frac{9}{8}\frac{\phi c^2 m^3}{(m^2 + m + 1)^2} \qquad (^1)$$

The equation for dp can therefore be written:

$$dp = \frac{\mu Q_m\, dr}{S k \rho}\left[1 + \frac{(m^4 - 1)\, c\, Q_m}{16\, n''\, S \mu\, \phi\, m}\right]$$

The quantity:

$$u = \frac{(m^4 - 1)\, c}{16\, n''\, \phi\, m}$$

(in practice we have $u \approx 10^{-3}$ CGS) homogeneous to a length will be called "shape parameter". In a limited range of flow rates it appears that a correlation can be established. The shape parameter u seems to "correlate" with the exponent n of Eq. 22.13 (see Fig. 22.33). We therefore will have:

$$dp = \frac{\mu Q_m}{S k \rho}\left(1 + \frac{u Q_m}{\mu S}\right) dr \qquad \text{(Eq. 22.32)}$$

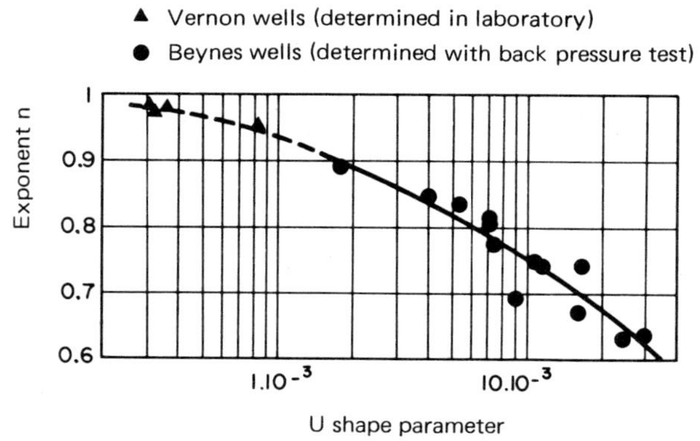

Fig. 22.33. Variation of n with u shape parameter.
(from E. Clarac, R. Monicard, L. Richard, 5th WPC, New York, 1959).

---

(1) If $m = 1$, i.e. for zero conicity, we have $k = \frac{\phi c^2}{8}$ as in the analogy with Poiseuille's law.

The expression above constitutes the elementary pressure drop law and it can be seen that it is a quadratic equation.

22.31 Integration of the quadratic equation in permanent flow: radial circular flow

Integration between the limits a and R in the case of radial circular flow for which $S = 2\pi rh$, leads immediately to:

$$\int_a^R \rho \, dp = \frac{\mu Q_m}{2\pi kh} \ln \frac{R}{a} \left[1 + \frac{u Q_m}{2\pi h \mu \ln \frac{R}{a}} \left(\frac{1}{a} - \frac{1}{R} \right) \right] \qquad \text{(Eq. 22.311)}$$

In the general case, R is much larger than a and the expression is simplified. In addition, if we note that $2\pi ha$ is the wall surface area of the borehole in the producing formation q_m the unit mass flow rate, can be introduced. We then have:

$$\int_a^R \rho \, dp = \frac{a\mu}{k} \cdot q_m \ln \frac{R}{a} \left(1 + \frac{u}{\mu \ln \frac{R}{a}} q_m \right) \qquad \text{(Eq. 22.312)}$$

In view of the practical values of u, which can be calculated, from μ and q_m, the second terme of the paranthesis is negligible before the unit for liquids but is not negligible for gases.

22.32. In parallel flow when integrating from 0 to 1 we have:

$$\int_0^1 \rho \, dp = \frac{\mu l Q_m}{Sk} \left(1 + \frac{u Q_m}{\mu S} \right) \qquad \text{(Eq. 22.321)}$$

and, with the introduction of the unit mass flow rate

$$\int_0^1 dp = \frac{\mu l}{k} q_m \left(1 + \frac{u}{\mu} q_m \right) \qquad \text{(Eq. 22.322)}$$

Analysis of losses of head for gaseous or liquid flows in cores therefore makes it possible to determine the shape parameter u and therefore to predetermine the quadratic law for circular flow.

22.33. The quadratic equation for pressure drops along the path d*r* is written:

$$\rho \frac{dp}{dr} = A Q_m + B Q_m^2 \qquad \text{(Eq. 22.331)}$$

where A and B are two numerical coefficients dependent upon the fluid, the porous medium and flow geometry.

In the case of the real flow of oil in a field it has been shown that the term BQ_m^2 is completely negligible before the term AQ_m while it has to be taken into consideration in the case of gas since the viscosities and flow rates usually found make it 500 times greater.

22.4. Conditions to be respected for assuring validity of Darcy's law

The generalized form of Darcy's law is written:

$$\rho \, dp = \frac{\mu}{k} \cdot \frac{Q_m}{S} \, dr \qquad \text{(Eq. 22.41)}$$

The elementary pressure drop law is written:

$$\rho \, dp = \frac{\mu Q_m}{sk} \left(1 + \frac{uQ_m}{\mu S}\right) dr \qquad \text{(Eq. 22.42)}$$

i.e. this is Darcy's law with a corrective term.

For Darcy's law to be valid it is necessary that the corrective term in Eq. 22.42 be negligible. If the limit is fixed at $\dfrac{uQ_m}{\mu S} < 0.01$ we can estimate a maximum rate of flow and a ΔP which is not to be exceeded for a given sample and fluid.

Let us suppose that there is a plug with the following characteristics:

$$\begin{cases} \phi = 10\,\% = 0.10 \\ m \approx 2\,n'' \text{ with } m = 3 \\ c = 2.10^{-4} \text{ cm} \\ S = 4 \text{ cm}^2 \end{cases}$$

We have:

$$\begin{cases} u = \dfrac{c\,(m^4 - 1)}{16\,n''\,\phi m} \# 2.2.10^{-3} \text{ CGS} \\ \dfrac{uQ_m}{\mu S} \# 3.1\,Q_m \end{cases}$$

with $\mu = 180.10^{-6}$ poise.

22.41. Case of air flow

$$\frac{uQ_m}{\mu S} < 0.01$$

if

$$Q_m < 3.2.10^{-3} \text{ g/s} \qquad \text{(Eq. 22.411)}$$

or

$$Q_a < 2.5 \text{ cm}^3/\text{s} \qquad \text{(Eq. 22.412)}$$

If Darcy's law is applied, it is possible to calculate the pressure gradient which must not be exceeded. According to Eq. 22.12 and 22.412 we have:

$$Q_a = \frac{Sk}{\mu L} \cdot \frac{P_G^2 - P_F^2}{2\,P_a} < 2.5$$

i.e.

$$P_G^2 - P_F^2 < \frac{5\,\mu L}{Sk} \cdot P_a$$

If

$$P_a = 1 \text{ atm} \qquad S = 4 \text{ cm}^2$$
$$\mu = 180.10^{-6} \text{ P} \qquad I = 2.5 \text{ cm}$$

then

$$P_G^2 - P_F^2 < \frac{56.9}{k}$$

with k expressed in millidarcys.

Hence the limits calculated for P_G (abs. bars) for P_F (abs. bars) and k (mD) :

P_F (abs. bars) \ k (mD)	0.1	1	10	100
0 (vacuum)	23.85	7.50	2.38	0.75
0.5	23.86	7.56	2.44	0.90
1 (atm)	23.88	7.61	2.59	1.26
2	23.94	7.80	3.11	2.14
5	24.37	9.05	5.54	5.06
10	25.86	12.53	10.28	10.03
50	55.40	50.56	50.05	50.006
100	102.80	100.27	100.03	100.004

22.42. Case of liquid flow

In the case of liquid flow with a viscosity which is much higher than that of gas the corrector term is much smaller.

In the preceding case, involving water where $\mu \approx 10^{-2}$ P, we have:

$$\frac{uQ_m}{\mu S} = 0.055\ Q_m$$

If the term is < 0.01, we have $Q_m < 0.2$ g/s.

Let $Q_{\text{water}} < 0.2\ \text{cm}^3/\text{s}$. Now the integration of Darcy's law for the preceding plug in the case of a liquid gives:

$$Q = \frac{Sk}{\mu} \cdot \frac{P_G - P_F}{L}$$

i.e.

$$\frac{Sk}{\mu} \cdot \frac{P_G - P_F}{L} < 0.2$$

where the maximum pressure gradient is

$$\frac{P_G - P_F}{L} < 0.2 \frac{\mu}{Sk}$$

or here

$$\frac{P_G - P_F}{L} < \frac{50}{k} \left(\frac{\text{bars}}{\text{cm}}\right)$$

where k is expressed in millidarcys and P in bars.

The table below gives the gradient values in bars/cm which must not be exceeded in relation to the k values (mD);

k(mD)	1	10	100	1 000
$\Delta P/L$ (bars/cm)	50	5	0.5	0.05

If the permeabilities are relatively high, excessively large pressure gradients should not be used or errors will appear in the results.

It should be noted that liquid permeabilities are not usually determined because of the special assembly which is necessary and the variations in liquid viscosity which occur with temperature changes.

23. AVERAGE PERMEABILITIES FOR SEVERAL LAYERS

23.1. Parallel layers

23.11. Parallel linear flow

We have (Fig. 23.111) n layers with permeability k_j each of which has a rate of flow Q_j subject to the same $\Delta P = P_G - P_F$:

$$Q_j = \frac{1\, h_j\, k_j}{\mu} \cdot \frac{\Delta P}{L}$$

$$Q = \sum_{j=1}^{n} Q_j = \frac{1}{L}\frac{\Delta P}{\mu} \sum_{j=1}^{n} h_j k_j = \frac{\Delta P}{\mu} \cdot \frac{1}{L} \left(\sum_{j=1}^{n}\right) h_j \,.\, \bar{k}_{av}$$

so that average permeability is

$$\bar{k}_{av} = \frac{\sum_{j=1}^{n} h_j\, k_j}{\sum_{j=1}^{n} h_j} \qquad \text{(Eq. 23.111)}$$

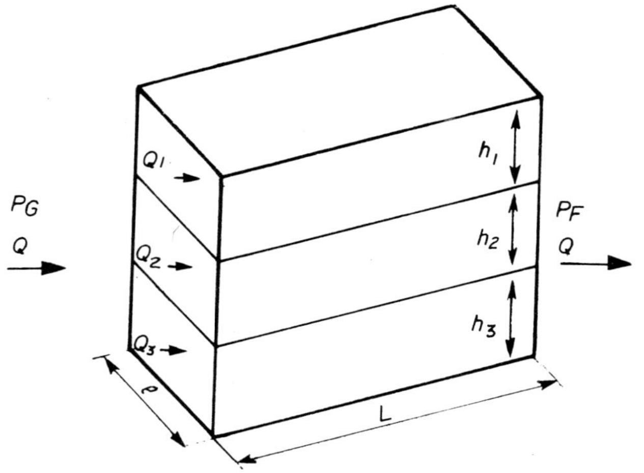

Fig. 23.111. Linear flow in parallel combination of beds.

23.12. Radial circular flow

This is the case of several superimposed layers flowing simultaneously in the well. Each layer supplies a rate of flow Q_i, determined by Darcy's law. The total rate of flow of the various layers is $Q = \Sigma Q_i$.

In the case shown in Fig. 23.121 of circular radial liquid flow we have:

$$Q_i = \frac{2\pi h_i k_i (P_R - P_a)}{\mu \ln \frac{R}{a}}$$

$$Q = \Sigma Q_i = \frac{2\pi h_t \bar{k}_{av} (P_R - P_a)}{\mu \ln \frac{R}{a}}$$

so that

$$\bar{k}_{av} = \frac{\sum_{j=1}^{n} h_i k_i}{h_t} \qquad \text{(Eq. 23.121)}$$

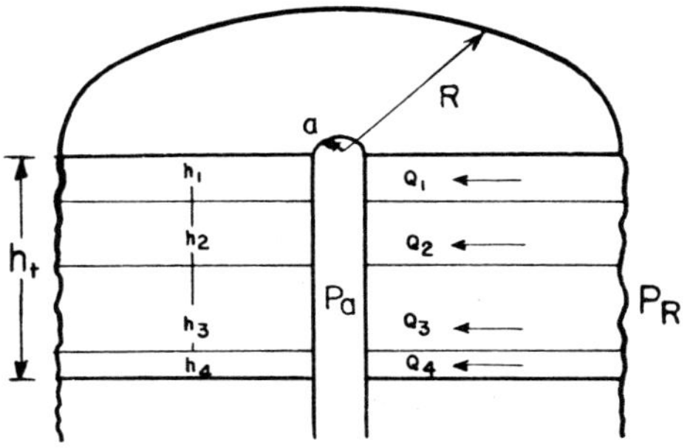

Fig. 23.121. Radial flow, parallel combination of beds.

This average permeability value can be calculated on the basis of permeability profiles measured during core analysis. This value can be compared with that obtained through other methods, i.e. well flow tests or pressure build-up tests.

23.2. Series of layers

23.21. Parallel flow

We have (Fig. 23.211):

In layer 1

$$Q = \frac{Sk_1}{\mu} \frac{\Delta P_1}{L_1}$$

In layer 2

$$Q = \frac{Sk_2}{\mu} \frac{\Delta P_2}{L_2}$$

In layer 3

$$Q = \frac{Sk_3}{\mu} \frac{\Delta P_3}{L_3}$$

etc.

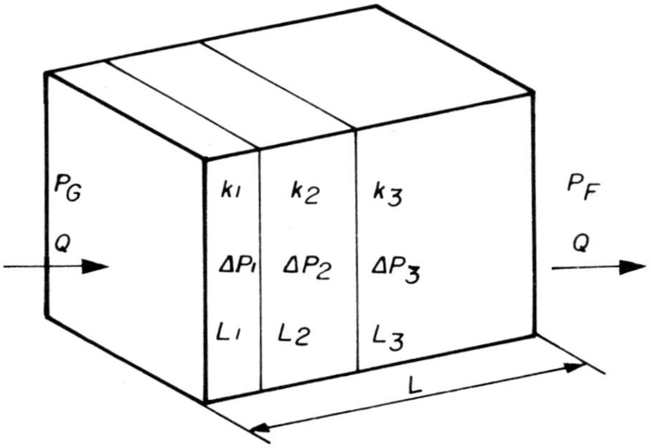

Fig. 23.211. Linear flows, series combination of beds.

The total is:

$$Q = \frac{S\overline{k}_{av}}{\mu} \cdot \frac{P_G - P_F}{L}$$

now

$$P_G - P_F = \sum_{j=1}^{n} \Delta P_j$$

so that $\overline{k}_{av}$

$$\overline{k} = \frac{L}{\sum_{j=1}^{n} \frac{L_j}{k_j}} \qquad \text{(Eq. 23.211)}$$

23.22. Radial circular flow

Permeability in the neighborhood of the well changes with the effects of well completion depending upon the method and treatment. From the practical point of view it is as if there were two layers in series and the average permeability of the group is then determined (Fig. 23.221).

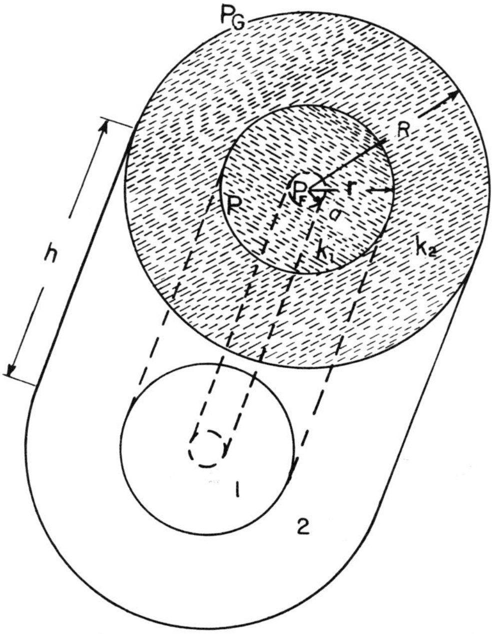

Fig. 23.221. Radial flow, series combination of beds.

When Darcy's law is applied we have:

In layer 1

$$Q = \frac{2\pi h k_1}{\mu} \; \frac{P - P_F}{\ln \frac{R}{a}}$$

In layer 2

$$Q = \frac{2\pi\, h k_2}{\mu} \frac{P_G - P_F}{\ln \frac{R}{r}}$$

and for (1 + 2)

$$Q = \frac{2\pi h \overline{k}_{av}}{\mu} \frac{P_G - P_F}{\ln \frac{R}{a}}$$

We also have:

$$P_G - P_F = (P_G - P) + (P - P_F)$$

so that

$$\overline{k}_{av} = \frac{\ln \frac{R}{2}}{\frac{1}{k_1} \ln \frac{r}{a} + \frac{1}{k_2} \ln \frac{R}{r}} \qquad \text{(Eq. 23.221)}$$

The calculation can be extended to a larger number of rings.

This calculation can be utilized for estimating the effects of mud invasion, acidizing or fracturing.

24. VARIATIONS IN PERMEABILITY IN THE CASE OF FLUID-MEDIUM REACTIONS

The injection of fresh water into a clayey sample gives zero permeability. The injection of fresh water into a reservoir leads to a reduction in permeability, especially if clay is present, even in very small quantities. Short tests during drilling very often show very low permeability values.

Figure 24.1 indicates the values of k measured for air, salt water and fresh water (according to laboratory measurements by *Core Laboratories Inc.*).

Figure 24.2 shows variations in water permeability according to salt content for various clay contents. It can be seen that it is necessary to reach a salt content of the order of 50 000 to 60 000 ppm for clay swelling to be reduced so that a reduction in permeability can be avoided.

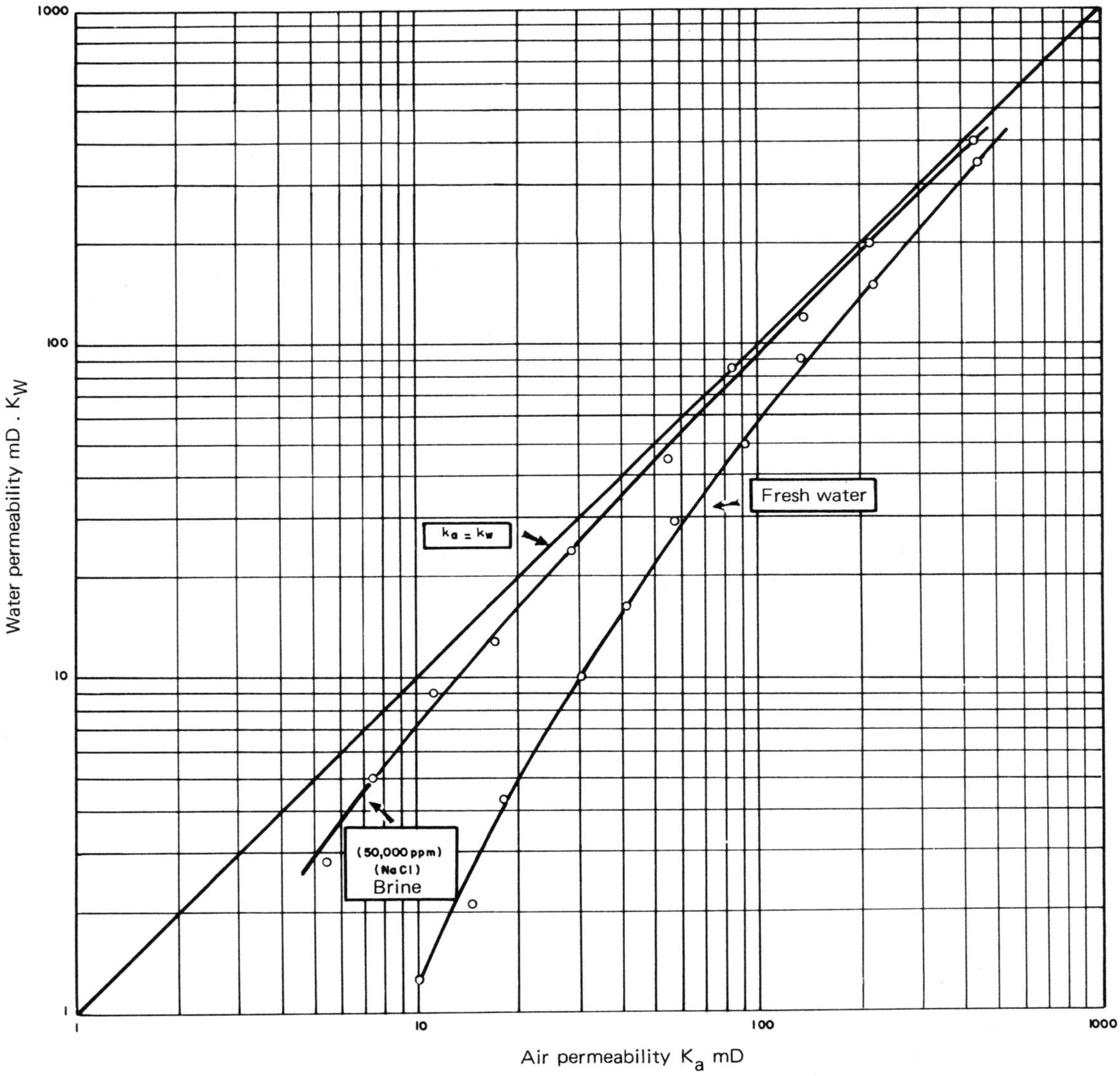

Fig. 24.1. Permeability to water versus permeability to air (from *Corelab*).

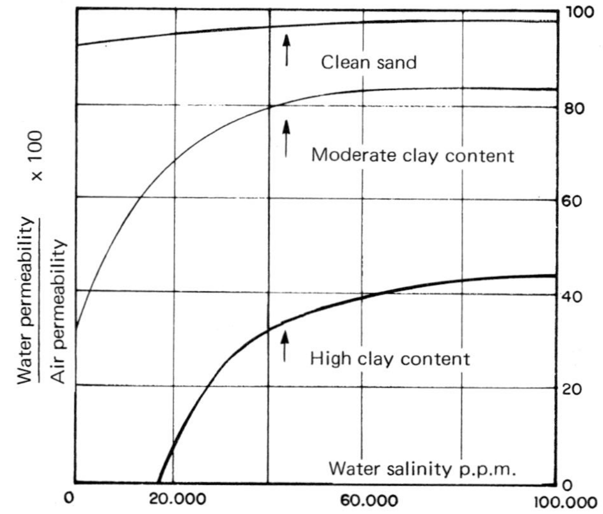

Fig. 24.2. Variations in water permeability with salinity and clay content.

25. EFFECT OF GEOSTATIC PRESSURE ON PERMEABILITY

When cores are brought up from the well bottom, all the convergent forces are eliminated and the rock matrix is allowed to expand in all directions. The shape of the channels for the flow of fluids inside of the core is therefore changed.

Compaction of the core due to geostatic pressure can produce a 25% increase in its permeability when it is taken out of the borehole (Fig. 25.1). Some cores

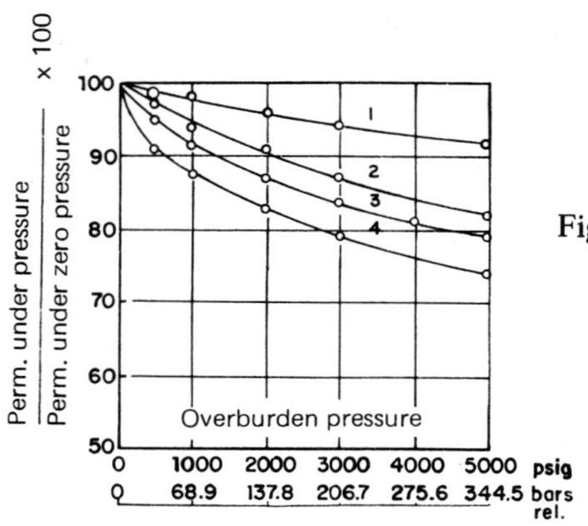

Fig. 25.1. Variations of k with overburden pressure.

1. Basal Tuscaloosa Mississippi.
2. Basal Tuscaloosa Mississippi.
3. Southern California Coast.
4. Los Angeles Basin.

(from Fatt and Davis).

are more compressible than others. Consequently, more data are necessary for developing the empirical correlations necessary for correcting the permeability value as measured at the surface for geostatic pressure.

Table 25.1 below (according to Le Tirant *et al.*, *Manuel de fracturation hydraulique,* Ref. 22) gives some very general values. It should be noted that fissured rock permeability drops rapidly when effective stress increases and then varies little.

TABLE 25.1

Type of rocks	Variations in permeability in terms of state of stress	$\frac{k_{reservoir}}{k_0}$
Consolidated sandstones without fissures	Small variation in $\frac{k}{k_0}$	0.7 to 0.9
Consolidated fissured sandstones	Very large variation in $\frac{k}{k_0}$	up to 0.05 to 0.10
Relatively unconsolidated sandstone (sometimes at great depths ($\geqslant$ 3 000 m)	Very large variation in $\frac{k}{k_0}$ Permeability drops very rapidly with increase in stress and then remains relatively constant	0.05 to 0.10
Organic detrital deposits	Continuous decrease in $\frac{k}{k_0}$	0.05 to 0.10

26. RELATION BETWEEN PERMEABILITY AND PORE MORPHOLOGY

We have previously seen that there is a possible analogy between Darcy's law and Poiseuille's law. The following relation between permeability k, pore radius r and porosity ϕ is arrived at:

$$k = \frac{\phi r^2}{8} \qquad \text{(Eq. 26.1)}$$

This relation is not a rigorous one but it does, however, make it possible to form a reasonable idea of pore dimensions.

Purcell tried to calculate permeability by considering the curve for mercury penetration into samples as an indication of the number of channels with radius r existing in the rock. The image is then that of a bundle of capillary tubes with different radii and with each radius existing for a certain number of tubes. He obtained:

$$k = \frac{1}{2} \phi T^2 \cos^2 \theta \int_0^1 \frac{ds}{P_c^2} \quad \text{(Eq. 26.2)}$$

where s is mercury saturation $= \dfrac{\text{Hg vol.}}{\text{pore volume}}$.

Pore radius and capillary pressure P_c are related by the equation:

$$P_c = \frac{2T}{r} \cdot \cos \theta$$

T = interface, θ = angle of contact.

The determination is not precise when this method is used. The author introduced a lithological factor F' making it possible to adapt the result to reality:

$$k = \frac{F'}{2} \cdot \phi T^2 \cdot \cos^2 \theta \int_0^1 \frac{ds}{P_c^2} \quad \text{(Eq. 26.3)}$$

As we shall see below, this lithological factor F' is merely the inverse of tortuosity.

In addition, in calculating elementary pressure drop by identifying Darcy's equation and that for the loss of head due to viscous friction, the following was assumed:

$$k = \frac{9}{8} \frac{\phi m^3 c^2}{(m^2 + m + 1)^2} \quad \text{(Eq. 26.4)}$$

The difficulty consists in determining m exactly. In practice it is possible to define c by the highest pressure and mc by the lowest according to the curve for capillary pressure resulting from mercury injection.

These results are quite satisfactory, but the method has not been used enough for a complete discussion to be possible.

27. EVALUATION FACTOR FOR PERMEABILITY WITH OTHER PARAMETERS: CONCEPT OF TORTUOSITY, THE COEFFICIENT OF KOZENY

According to the analogy between the equations of Darcy and Poiseuille, the permeability of a capillary tube with the radius r is:

$$k = \frac{r^2}{8}$$

If it is admitted that a porous medium is equivalent to a network of capillary tubes, the permeability of the medium can be calculated and depends upon pore porosity and distribution. A capillary network will resemble parallel layers with different permeabilities (see Fig. 23.111). The average permeability $\bar{k}$ of the group will be calculated by means of Eq. 23.111:

$$\bar{k} = \frac{\sum_{j=1}^{m} k_j \cdot A_j}{\sum_{j=1} A_j} \phi \qquad \text{(Eq. 27.1)}$$

where k_j is the permeability of a capillary tube with the cross section A_j/n_j. The quantities k_j and A_j can be determined in terms of the tube radius. Hence we have:

$$k_j = \frac{rj^2}{8}$$

$$A_j = n_j \pi r j^2$$

$$\sum_{j=1}^{m} A_j = \phi \cdot A$$

where

n_j = the total number of tubes with radius r_j,
ϕ = porosity of the medium,
A = cross section of capillary tube network.

If we replace k_j and A_j in terms of r_j in Eq. 27.1 we obtain:

$$\bar{k} = \frac{\phi}{8} \frac{\sum_{j=1}^{m} n_j r_j^4}{\sum_{j=1}^{m} n_j r_j^2} \qquad \text{(Eq. 27.2)}$$

It should be noted that the permeability of a tube network is not only a function of pore volume but of system configuration as well.

If the system consists of a network of n similar capillary tubes, Eq. 27.2 is written:

$$\bar{k} = \frac{\phi r^2}{8} \qquad \text{(Eq. 27.3)}$$

The total internal surface area S_i per pore unit volume can be defined in terms of capillary radius and we have:

$$S_i = \frac{2}{r} \qquad \text{(Eq. 27.4)}$$

The combination of Eq. 27.3 and 27.4 gives $\bar{k}\,(S_i)$, i.e.

$$\bar{k} = \frac{1}{2} \cdot \frac{\phi}{S_i^2} \qquad \text{(Eq. 27.5)}$$

If in Eq. 27.5 the coefficient constant 1/2 is replaced by $\frac{1}{k_z}$, the expression obtained is Kozeny's equation:

$$\bar{k} = \frac{\phi}{k_z \cdot S_i^2} \qquad \text{(Eq. 27.6)}$$

Rose and Willie have used this equation. It is clear that for a porous medium when, for example, the formation factor is measured, the path followed by the electrons is approximately the same as that of the moving particles. The liquid molecules do not move along the shortest path fom one face to another of the porous sample. The permeability results for this flow network are given by Willie's equation below:

$$k = \frac{\phi}{k_o\, S_i^2} \cdot \left(\frac{L}{L_a}\right)^2 \qquad \text{(Eq. 27.7)}$$

where

k = permeability of porous medium,
k_o = shape factor,
L = length of sample,
L_a = length of flow path,
$\left(\frac{L_a}{L}\right)^2 \approx 4$ (according to Birks).

If we call $\left(\frac{L_a}{L}\right)^2 = \tau$ the tortuosity of the porous medium, we have:

$$k_z = k_o \tau$$

the constant of Kozeny.

The two Eq. 27.6 and 27.7 are thus identical.

According to Willie and Spangler, the coefficient of Kozeny can be calculated on the basis of the saturation-capillary pressure curve.

In reality, the inverse of tortuosity is Purcell's lithological factor F', Eq. 26.3.

The relation between k and ϕ is determined theoretically if another variable is known. The specific surface, porosity and permeability vary considerably in a porous medium.

28. PHYSICAL ANALOGIES WITH DARCY'S LAW

Darcy's law (parallel flow). We have:

$$Q = \frac{k}{\mu} \cdot S \cdot \frac{\Delta P}{\Delta L}$$

28.1. Ohm's law

We have:

$$I = \frac{U}{R}$$

where

I = current (amperes),

U = potential difference (volts),

R = resistance (ohms).

By introducing resistivity ρ ($\Omega \cdot$ cm) $= \frac{1}{\sigma}$ conductivity, Ohm's law becomes:

$$I = \frac{AU}{\rho L}$$

(where A = section of conductor of length L).

When the above equations are compared, the following analogies are obtained:

$$Q \approx I \qquad \frac{k}{\mu} \approx \frac{1}{\rho} \qquad \frac{\Delta P}{\Delta L} \approx \frac{U}{L} \qquad \text{(Eq. 28.11)}$$

28.2. Fourier's equation for heat transmission is written:

$$q = k' \cdot A \cdot \frac{\Delta T}{L}$$

where

q = quantity of heat per unit of time,
A = section of conductor,
L = length,
ΔT = temperature variation,
k' = thermal conductivity.

When comparison is made with Darcy's law the following analogies are obtained:

$$Q \sim q \qquad \frac{k}{\mu} \sim k' \qquad \frac{\Delta P}{\Delta L} \sim \frac{\Delta T}{L} \qquad \text{(Eq. 28.21)}$$

These analogies are often utilized for experimentally solving certain production or injection problems on scale models.

29. LABORATORY MEASUREMENT OF RESERVOIR ROCK PERMEABILITY WITH SINGLE-PHASE FLUIDS, PERMEAMETER

The technique for laboratory measurements of rock sample permeability varies depending upon:

(a) The state of consolidation of the samples.
(b) Sample dimensions.
(c) The fluid used (gas-water-oil).
(d) The range of pressures on the sample.

For consolidated soils, samples are generally collected in geometric shapes:

(a) Either full cylinders with the following approximate dimensions: diameter = 2.3 cm to 4.0 cm and height = 2.5 cm to 5.0 cm.
(b) Or cubes with 2 cm sides.
(c) Or hollow cylinders for radial circular flow measurements.

In certain cases when the state of consolidation is insufficient, it is necessary to make a small assembly or to coat:

(a) For example to coat a fragment with wax in order to measure permeability in side wall core analysis or for unconsolidated sand.

(b) Coating with bakelite or plastic ("plasticore" by means of a thermo-setting plastic).

A special technique is necessary for measuring the permeability of cores from fissured limestone deposits. These measurements are made on whole cores (the Whole Core Analysis technique described in Chapter 5 of this course).

29.1. Measurement of air permeability k_a for geometric samples (with parallel flow)

The instruments which are currently utilized make possible measurements of 1.10^{-4} to 20 darcys.

Two types of instruments are usually used in the laboratory:

(a) Variable head permeameter, IFP type.
(b) Constant head permeameter, *Core Laboratories* type.

29.11. Variable head permeameter, IFP type (derived from the instrument of M. Métrot)

In Fig. 29.111 we have:

(a) Core holder (Fig. 29.112) with a rubber stopper.

(b) A glass tube for flow rate measurements. This tube consists of three different sections with gradations from the top down of 1-10-50 so that there are three ranges of sensitivity (for $1 < k_a < 2\,000$ mD).

(c) A constant level water tank.

(d) A rubber bulb for bringing water up to desired point (or for suction under partial constant vacuum).

(e) A valve for cutting off the suction bulb from the inner space.

This instrument is described at length in the *Manuel de Travaux Pratiques* where the theory is presented in detail. Permeability is given by an equation with the following form:

$$k_a = \frac{B\mu}{\Delta t} \cdot \frac{1}{S} = \frac{\text{Const.} \cdot L}{\Delta t}$$

where

B = a constant depending upon the geometry and units of the apparatus,
μ = air viscosity, a function of temperature and humidity,
1 = sample length,
S = sample section,
Δt = flow time between two level marks.

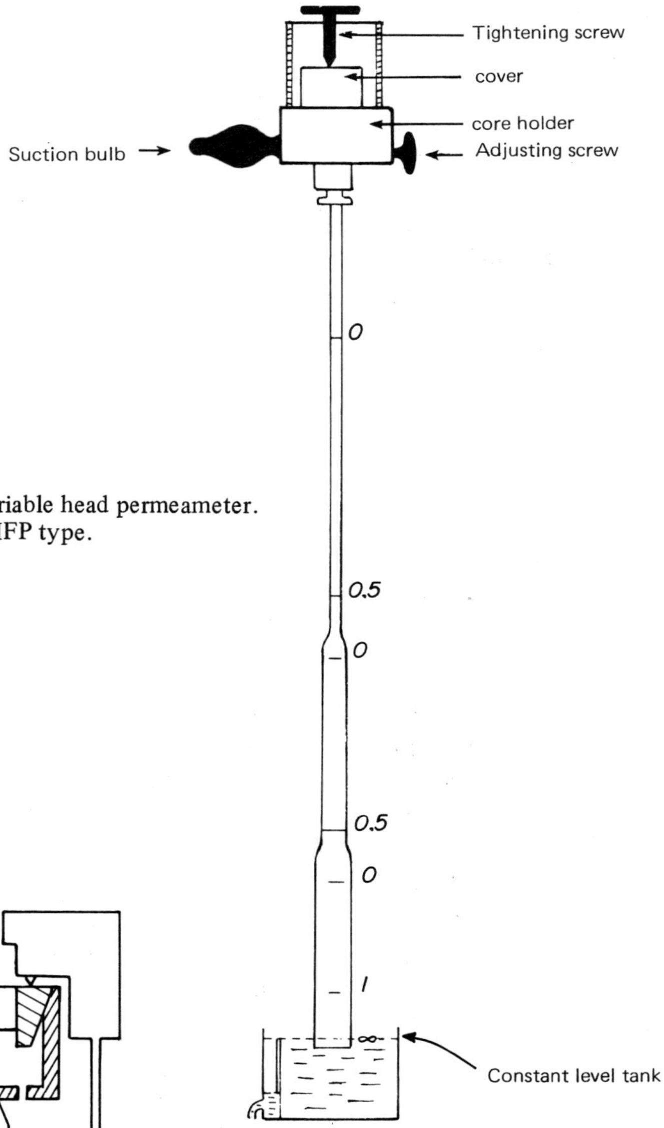

Fig. 29.111. Variable head permeameter. IFP type.

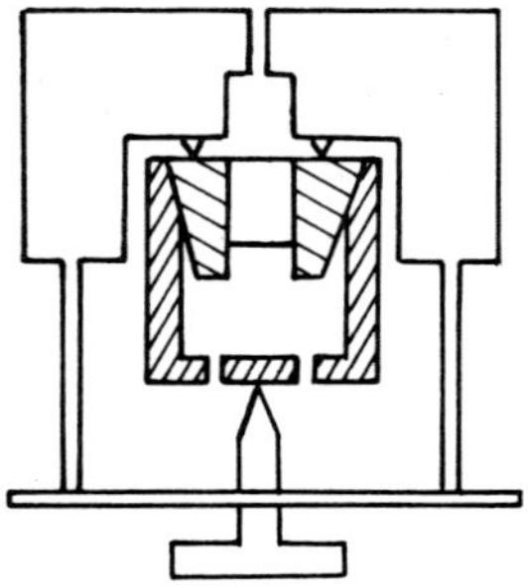

Fig. 29.112. Core holder of changeable load air permeameter.

29.12. Constant head permeameter, Core Laboratories type

Figure 29.121 gives a diagram of this apparatus:

(a) Upstream and downstream pressures are indicated by manometers (gauges or mercury manometers).

(b) Gas flow is measured by means of a calibrated outlet.

Air permeability, k_a, is given by:

$$k_a = \frac{2\mu \cdot Q_a \cdot L}{S} \frac{P_a}{(P_1^2 - P_2^2)} \qquad \text{(Eq. 29.121)}$$

Q_a = rate of air flow measured at pressure Pa (atmospheric),
L = sample length,
S = sample section,
P_1 = pressure upstream,
P_2 = pressure downstream,
μ = air viscosity.

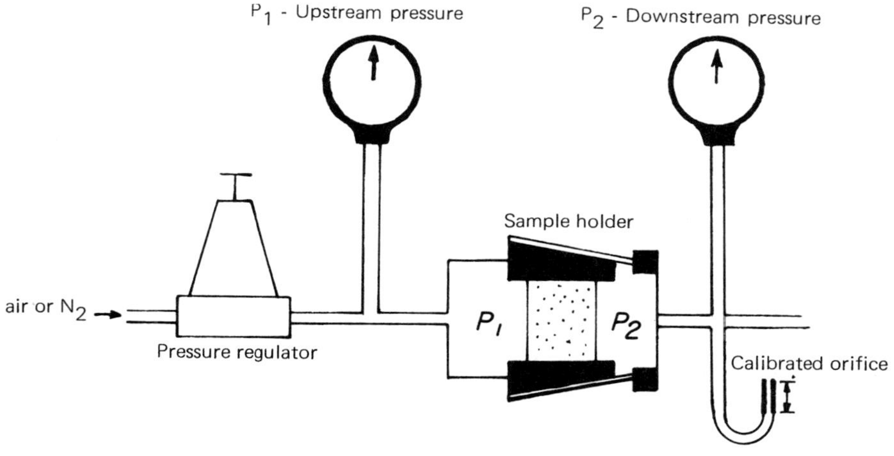

Fig. 29.121. Permeameter : schematic flow diagram.

Remark:

Practical units:

$$k_a \text{ (mD)} = \frac{2\,\mu\,(cP)\;.\;Q_a\,(\text{cm}^3/\text{s}\;.\;L\,(\text{cm})\,P_a\,(\text{abs. atm})\;.\;1\,000}{(P_1^2 - P_2^2)\,(\text{abs. atm})^2\;.\;S\,(\text{cm}^2)}$$

(Eq. 29.122)

29.2. Measurement of gas permeability for hollow cylindrical samples: dry gas

Figure 29.21 gives a diagram of the apparatus utilized.

Measurements are made for circular radial flow. According to Darcy's law we have:

$$k = \frac{\mu Q_a}{\pi h} \ln \frac{R}{a} \cdot \frac{P_0}{P_1^2 - P_0^2} \qquad \text{(Eq. 29.21)}$$

k = permeability,
μ = gas viscosity,
Q_a = gas flow rate measured at pressure P_0,
R = outside radius of sample,
a = inside radius of sample,
h = height of sample,
ln = neperian log.

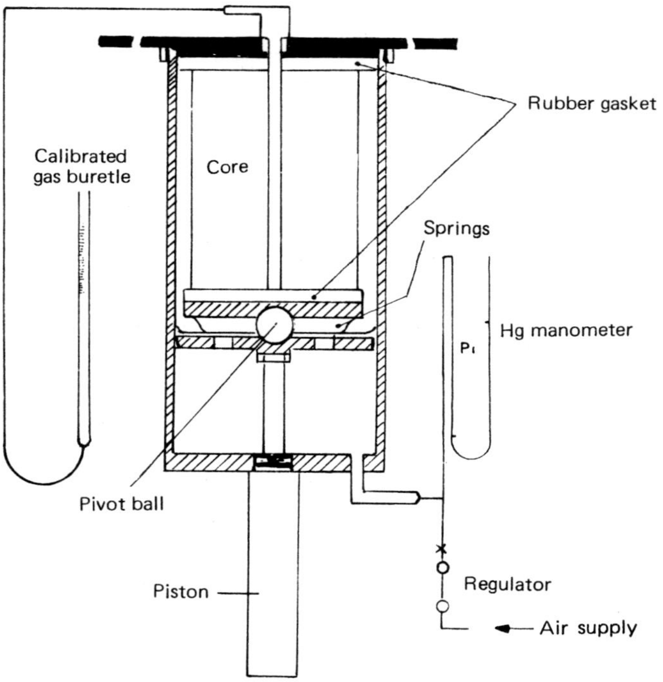

Fig. 29.21. Full diameter radial permeameter.

With practical units:

$$k_{(\text{mD})} = \frac{\mu\,(cP) \cdot Q\,(\text{cm}^3/\text{s}) \cdot \ln \frac{R}{a} \cdot P_0\,(\text{abs. atm}) \cdot 1000}{\pi \,.\, h(\text{cm}) \,\cdot\, (P_1^2 - P_0^2)\,(\text{abs. atm})^2} \qquad \text{(Eq. 29.22)}$$

The difficulties lie in the preparation of the samples.

29.3. Permeability measurements for large samples (air permeability)

In certain cases permeability is measured on large samples which may sometimes be cracked. *Core Laboratories Inc.* uses a permeameter such as the one shown in Fig. 29.31. For measuring horizontal permeabilities the cores are treated as follows:

(a) The ends are emersed in wax (Fig. 29.32) and a height G is left uncovered.
(b) Two screens (sieve type) are placed between the core and the rubber half-shell cylinders for allowing passage of the air flow (Fig. 29.33).
(c) The core and the rubber shells are gripped by a hydraulic system.
(d) Q_{air} is measured by means of the calibrated outlet.

Permeability is given by the relation:

$$k_a = C\,(SF)\,Q_a \qquad \text{(Eq. 29.31)}$$

with

$(SF) = 1.2\sqrt{LG}$; this factor depends upon core dimensions and screen opening,
C = Constant depending upon ΔP and the viscosity of the gas used.

Permeability is measured in the two perpendicular directions.

29.4. Utilization of the Hassler assembly

This assembly is an improvement on the use of rubber plugs whose tightness is limited at certain pressures. The core is placed in a flexible rubber tube (Fig. 29.41). At each end of the core there are steel fittings which make it possible to inject or collect fluids or to measure the pressure. Tightness is assured by inflation either with air or with a liquid.

This method, which may be less rapid than the plug, has undeniable advantages:

(a) Excellent tightness.
(b) Can be used for samples of very different sizes.
(c) Much higher pressures or ΔP can be used: it is possible to reproduce geostatic pressures.

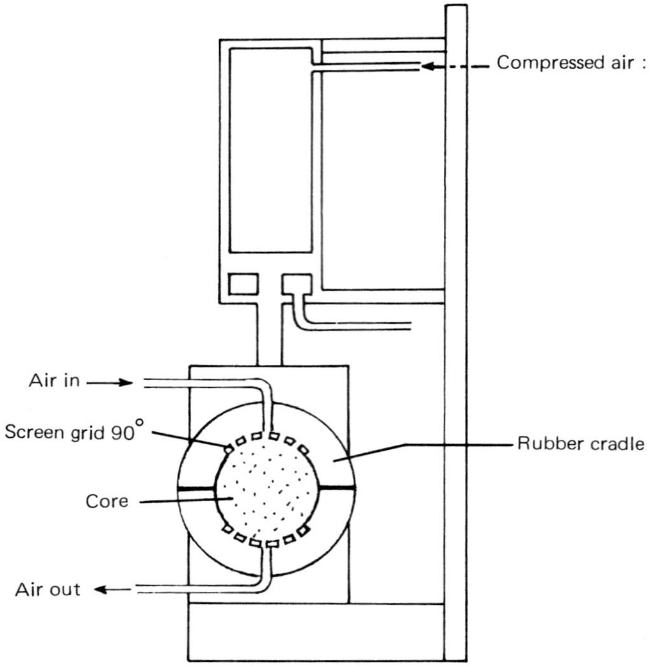

Fig. 29.31. Whole core permeameter.
(from *Core Laboratories, Inc.*).

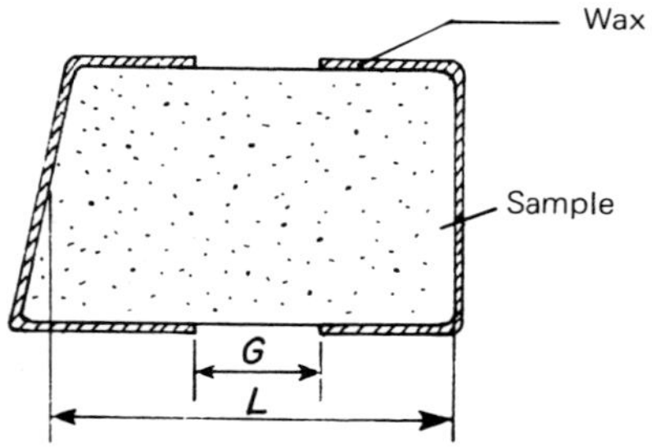

Fig. 29.32. Preparation of whole core for k_a measurements.

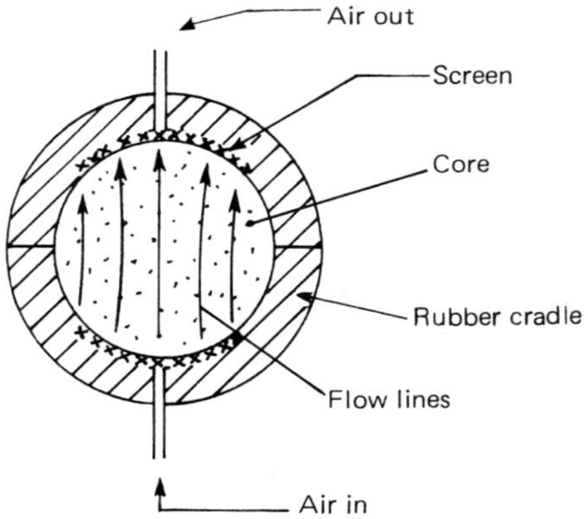

Fig. 29.33. Passage of air in sample during measurement.

(d) Permeability measurements in two directions H and V (Fig. 29.41).

(e) Measurement of air, oil or water permeability.

(f) Hassler assemblies can be used for various measurements such as relative water/oil, gas/oil permeabilities.

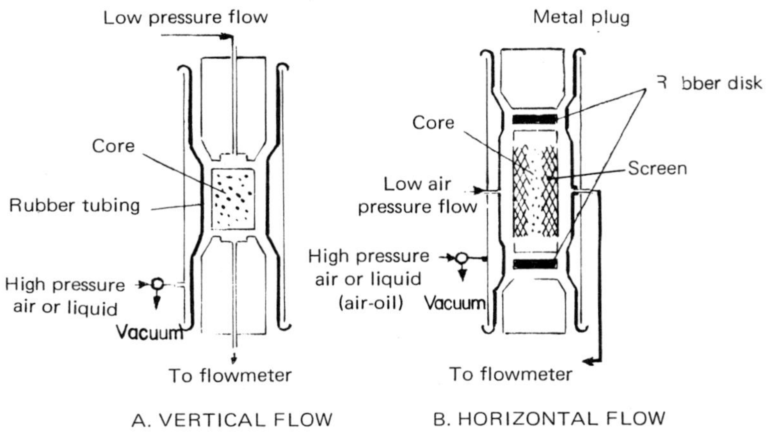

Fig. 29.41. Hassler type permeameter.

29.5. Permeability of unconsolidated media

It is possible to measure the permeability of sands:

(a) A sample is collected and is consolidated in some ways i.e. by means of paraffin, wax, plasticore etc.

(b) A permeameter with relatively unconsolidated sand is used. The sand is placed in a tube in a special way and the packing is very important (packing in waterpacking in air) as well as the compaction and the materials acting as cement. The permeability values vary during the measurement and become stabilised at the end of a certain period of time. (Fig. 29.51).

(c) In the case of the so-called side wall core analysis, a piece of sample is placed in the centre of a copper cylinder with the unit being placed on fine sand (Fig. 29.52). Wax is poured between the sample and the cylinder. After the base has been removed, the unit is placed in an air permeameter either at variable or constant head.

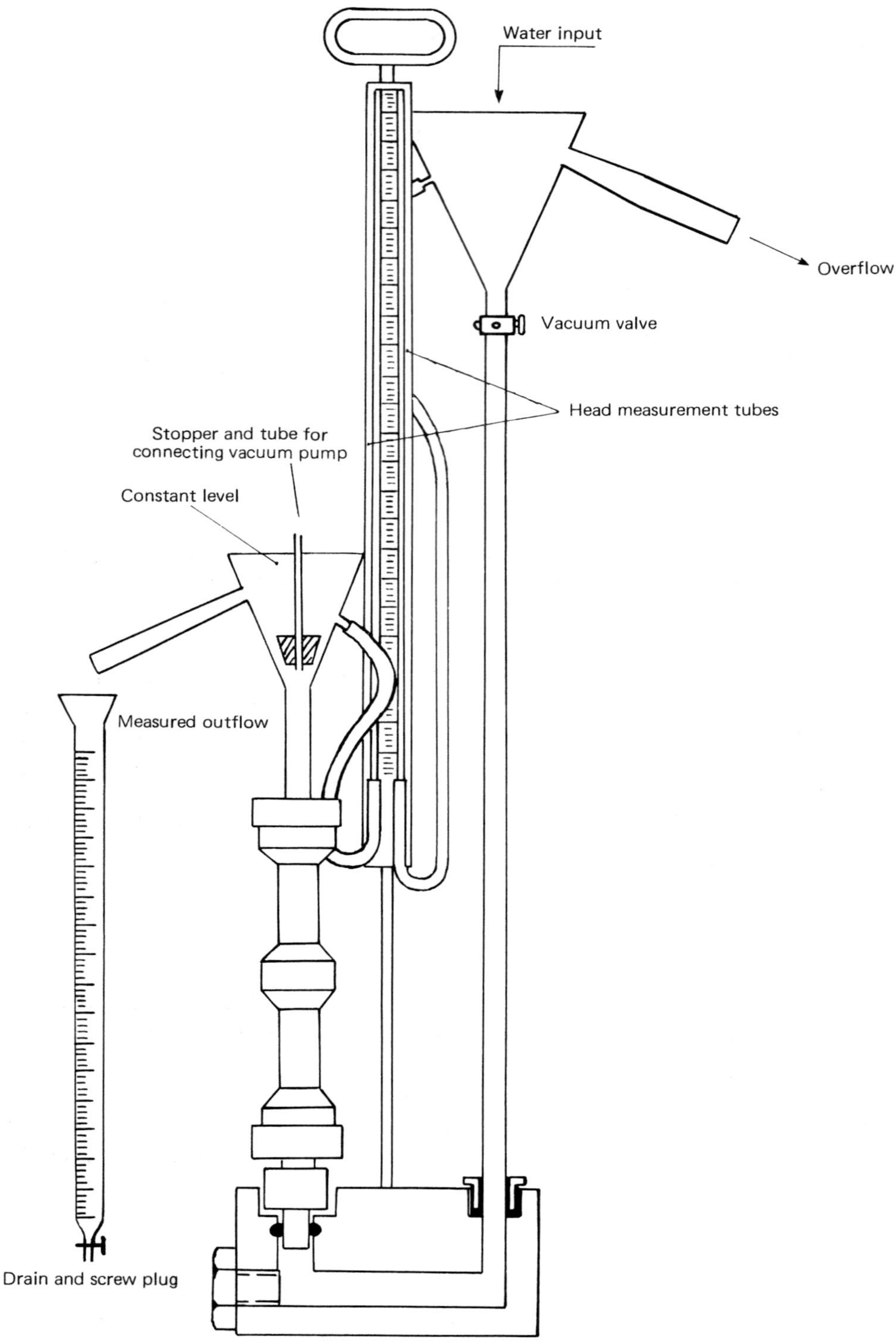

Fig. 29.51. Sand permeameter.
Overall view. IFP model.

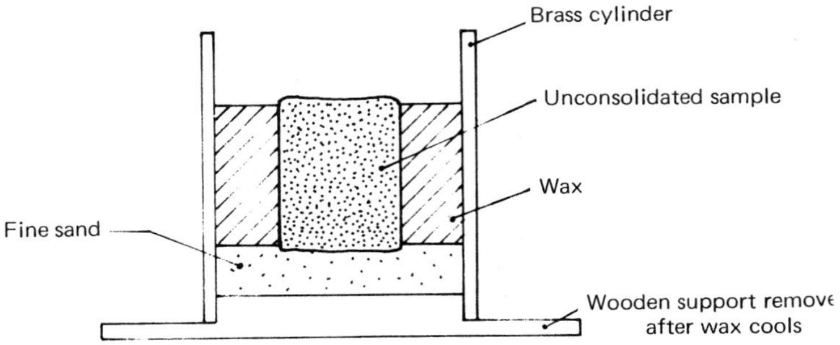

Fig. 29.52. Preparation of unconsolidated sample for k_a measurements.

2.10. INDIRECT DETERMINATION OF PERMEABILITIES AND CORRELATIONS BETWEEN POROSITY AND PERMEABILITY

The direct determination of permeabilities (based on Darcy's equation) supposes that a reservoir sample has been obtained in which a relatively consolidated piece with a geometric shape has been selected (cube or cylinder known as a plug). This small sample represents only a very small part of the reservoir and very often it is not a core. Hence indirect methods or empirical correlations are utilized in order to attempt to estimate reservoir permeability. The following methods can be noted:

2.10.1 Thin section porosity

Determination of thin section porosities by means of the injection of a synthetic colored resin under high pressure following establishment of a vacuum. An overall result for the sample is obtained which is an average value sand this is interesting but could often be insufficient. A representation of the real rock networks is obtained on the basis of which it is possible to calculate an "equivalent diameter" for the pores. By assimilating porous media to assemblies of cylindrical tubes with the radius r, we have:

$$K = \text{Const.} \cdot r^2$$

or

$$K = \text{Const.} \cdot \phi\, r^2$$

The constant varies depending upon the arrangements chosen.

2.10.2. Estimate of permeability by means of logs

Authors (Kozeny and Jacquin) studied the relation between porosity and permeability.

For a Fontainebleau sanstone Jacquin found an equation with the form:

$$k = \alpha \phi^{2m+1}$$

with

$$\text{mD} = \%$$

m = 1.75 cementation factor,
α = 3.10^{-2} constant for grains of given shape.

From the work of Kozeny, Wyllie and Rose, Schlumberger deduced an empirical form:

$$k^{1/2} = 250 \frac{\phi^3}{S_{wi}} \quad (\phi, S_{wi} = \text{fraction})$$

S_{wi} = irreducible water saturation or water saturation higher than transition zone.

A chart makes it possible to solve the above equation (ϕ and S_{wi} are decuded from logs).

Hence for sandstones with the same porosity an equation of the following form is used:

$$k^{0.3} = \frac{\text{Const.}}{S_{wi}}$$

These empirical formulae are very approximate and must be used with care. It is always a complex matter to go from permeability measurements to the potential rate of flow of a well. The latter can be established relatively exactly only through flow tests by formation testing.

2.10.3. Calculation of permeability on the basis of pore distribution (See Chapter 27)

Many approximate formulae have been given based upon the correlation of Darcy's law and Poiseuille's law. For example, for a homogeneous bundle of cylindrical tubes we have:

$$K_{\text{darcy}} = 0.127 \phi r^2$$

r = radius of tube in μ,
ϕ = porosity (fraction).

When capillary pressure curves related to pore distribution are introduced a correlation of the following form is used (Birks):

$$k_{\text{darcy}} = \frac{0.127 \sum_{0}^{\phi} r^2 \, d\phi}{(L_a/L)^2}$$

r = equivalent pore radius in microns = $\dfrac{2T\cos\theta}{P}$

ϕ = porosity (fraction),

$(L_a/L)^2$ = tortuosity of capillaries, average value = 4.

2.10.4. Correlation between internal surface areas of porous media and permeability (according to Bureau of Mines – Continental Oil Co. – Shell Oil Co.)

Kozeny's equation can be written as follows:

$$k = \frac{\phi^3}{5.0\,S^2\,(1-\phi)^2}$$

S = specific surface area.

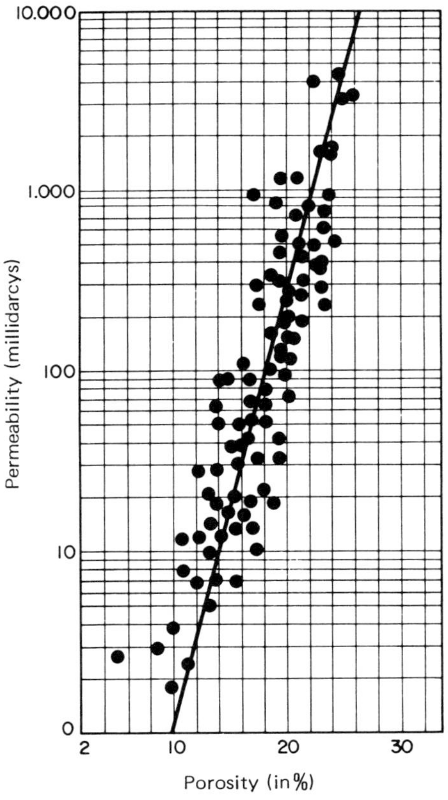

Fig. 2.10.51. Cazaux, Albian Field. Porosity-permeability correlation.
(from J.P. Dupuis, G. Oswald, J. Sens)

2.10.5. Permeability-porosity correlations

"The quantitative relations between porosity and permeability are obscure and variable" (Levorsen).

For field study purposes (reservoir engineering) a correlation of the following form is determined in the case of reservoirs belonging to the same formation:

$$\phi = \mathrm{a} \log k + b$$

where a, b are constants.

Figure 2.10.51 gives an example of the correlation $k - \phi$. There are sometimes two line segments.

A correlation between $k_{\text{horizontal}}$ and k_{vertical} of the following form is also developed:

$$K_H = a\, K_V^n$$

The vertical permeability of the reservoir plays a determining role in the effectiveness of gas-cap drive.

3

fluid saturation of reservoir rocks and capillary properties

31. FLUIDS CONTAINED IN RESERVOIR ROCKS AND INTERSTITIAL WATER

31.1. Fluids contained in the reservoir

As a result of the origins of the oil and its formation and migration conditions, the reservoir rocks contain the following fluids:

(a) Liquid hydrocarbons: oil from the light fraction to asphalts,
(b) Gaseous hydrocarbons.
(c) Water (salt water).

These fluids which are distributed in a certain manner in the porous medium under reservoir temperature and pressure conditions are, in general, found to have quite different distributions in the cores brought to the surface.

These modifications are due to the following factors:

(a) Firstly, to causes which are difficult to avoid:
. Invasion of drilling mud or filtrate.
. Gas expansion due to the fall in pressure during the raising of the core.

(b) Secondly, there are often handling errors such as the washing of the cores in water, or drying at high temperatures or the lack of preservation.

The quantity of fluid contained in the pores, expressed as a percentage of V_p **is called fluid saturation.**

31.2. Interstitial water. Definition

Interstitial or connate water is present in all oil or gas reservoirs. This water surrounds the grains and fills the small pores. In general, the hydrocarbons occupy the centre of the large pores and cracks. (Fig. 31.21).

Water saturation depends upon:

(a) Pore size and distribution.
(b) The height of the sample above the water-bearing zone.

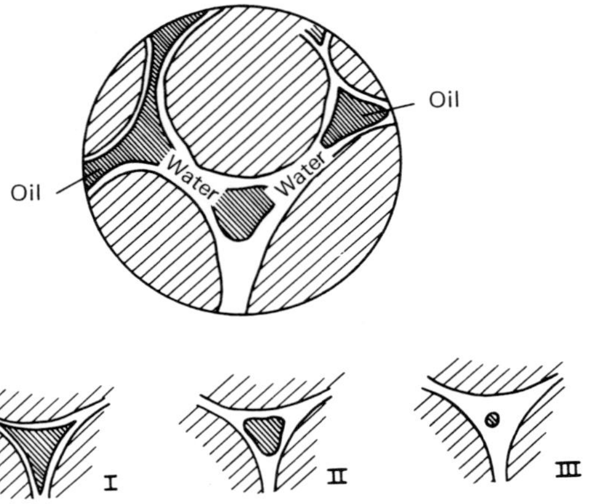

Fig. 31.21. Diagram showing distribution of oil nonwetting fluid (in black). Inside pores of a rock filled with water (wetting liquid).

1. Oil saturation of the order of 80% (productive zone).
2. Oil saturation of the order of 50% (transitional zone).
3. Oil saturation of the order of 10-20% (water-bearing zone).

Coarse grained sandstones, oolithic and vuggy limestones and all rocks with large pores have relatively low connate water saturation, while very fine grained sandstones have relatively high interstitial water saturation.

Water saturations decrease as the distance from the water-bearing zone increases.

Interstitial water saturation is determined at each elevation by the capillary attraction phenomena of the small pores.

32. CAPILLARY PROPERTIES OF ROCKS

The specific surface area (1) of rocks (boundary surface area of empty spaces in the porous medium) increases as the size of the constituent grains falls, as is shown in the table below:

Grain diameter (mm)	Surface area exposed to fluid (m^2/m^3 of total rock volume)
1.661	2 510
0.883	5 000
0.417	10 050
0.208	20 000
0.104	42 700
0.050	83 000

The role of surface phenomena in porous media increases in importance as the media become finer for a given pair of fluids.

These phenomena are linked to the relative solid surface affinities for each of these fluids and the interfacial tension between the fluids.

Some important ideas are reviewed below.

32.1. Review of some ideas concerning surface phenomena

32.11. Fluid/fluid interface and interfacial tension. Capillary pressure

Surface phenomena are due to particular molecular attractions which appear when two fluids which are not mixable in all proportions are in contact. If a drop of mercury is examined on a table, it can be observed to act as if an elastic membrane were stretched over the surface of the drop holding in the fluid. If we suppose an incision to be made in this membrane, it would be necessary to exercise a force perpendicular to the direction of the incision and proportional to its length in order to keep the two sides of the incision in contact. This force per unit of length is the **surface tension** of mercury T (T_{Hg} = 480 dyn/cm).

(1) Determination of the specific surface area is based upon the low temperature absorption phenomenon. It is deduced from the quantity of gas or vapor necessary for forming a monomolecular gas layer on the sample surface. The area occupied by each gas molecule is known.

If the other fluid is not air, the tension is called **interfacial.** The following values should be known:

(a) Surface tension of water = 73 dyn/cm.
(b) Water/oil interfacial tension ≈ 28 - 30 dyn/cm.

In most cases the contact surface between two fluids has a certain curvature except if this surface is very extended. There are two principal radii of curvature R_1 and R_2 at any given point. The fact that this curvature is not null leads to a difference in pressure between the fluids in contact.

This pressure difference is called capillary pressure, and the higher pressure is on the side where the surface curvature centre is located. The equation for surface equilibrium is as follows:

$$P_c = T\left(\frac{1}{R_1} + \frac{1}{R_2}\right) \qquad \text{(Eq. 32.111)}$$

Remark:

Interfacial tension exists even if $P_c = 0$.

32.12. Solid-interface contact and angle of contact

Experiments have shown (Figs. 32.121 and 32.122) that separation surfaces or menisci between two non mixable fluids touch the solid walls which bound them at a certain angle which is called the "angle of contact" of the two-fluid system and the solid. The angle of contact is in direct relation to the idea of surface wettability.

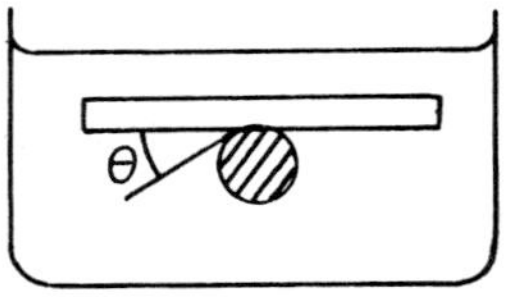

Fig. 32.121.

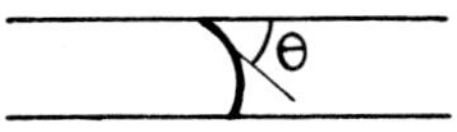

Fig. 32.122.

Water wets glass (it leaves a liquid trace) but mercury does not wet glass. The angle of contact for water is less than 90° while for mercury it is more than 90° (of the order of 140°).

This observation that the angle of contact is less than 90° could characterize the phase of preferential wetting of the solid surface (Figs. 32.123, 32.124).

In water near the menisci, the pressure is below the pressure in air by the quantity $h\rho_w g$ with the specific gravity of the air ρ_a not being taken into account:

$$P_c = h\rho_w g \qquad \text{(Eq. 32.120)}$$

h = capillary height,
ρ_w = specific gravity of water,
g = gravity constant.

On the other hand, in mercury (Fig. 32.124) near the meniscus, the pressure is above the pressure in air by the quantity $h \cdot \rho_{Hg} \cdot g$.

As was said above, the higher pressure is always on the side of the centre of curvature.

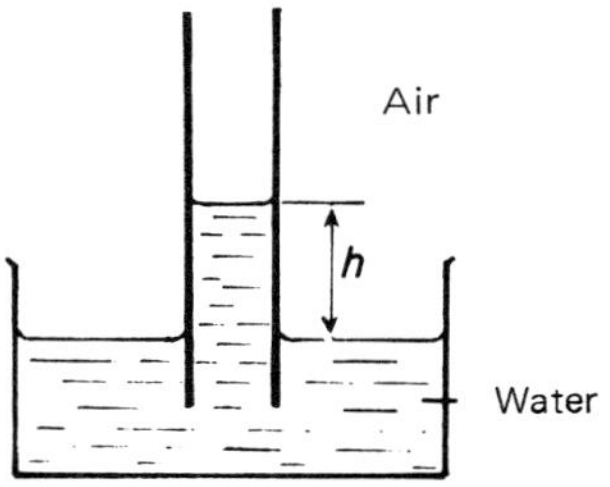

Fig. 32.123.

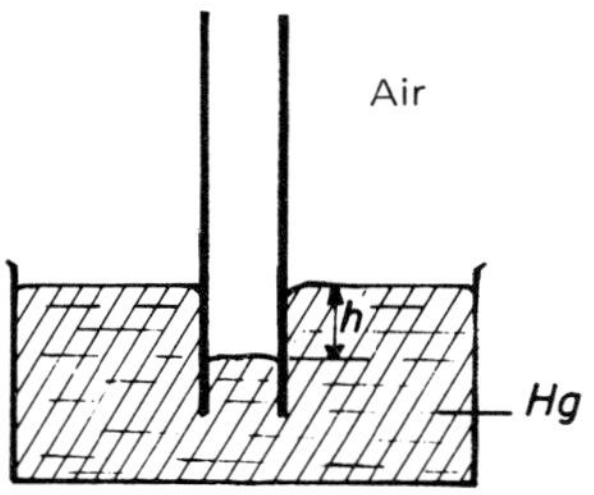

Fig. 32.124.

In general, the menisci are small and can be assimilated to spherical surfaces. Hence (Fig. 32.125):

$$P_c = \frac{2T}{R} \qquad \text{(Eq. 32.121)}$$

since $R = R_1 = R_2$.

There will therefore be the following equation for capillary pressure, angle of contact, interfacial tension and pore radius:

$$P_c = \frac{2T \cdot \cos\theta}{r} \qquad \text{(Eq. 32.122)}$$

Group $(T \cdot \cos\theta)$ characterizes the pair of fluids (T) and the solid (θ). In order to make the nonwetting fluid penetrate into pores with radius r with system (T, θ), a pressure P_c is necessary, while with system (T', θ') a pressure P'_c is necessary. Pressures P_c and P'_c are linked by the equation:

$$\frac{P_c}{T\cos\theta} = \frac{P'_c}{T'\cos\theta'} \qquad \text{(Eq. 32.123)}$$

This makes it possible to choose one pair of fluids or another in order to study pore morphology or saturation states corresponding to various values for capillary pressure.

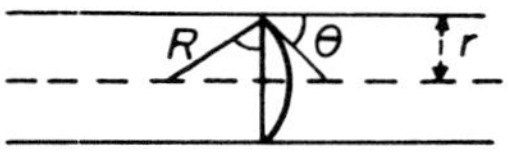

Fig. 32.125.

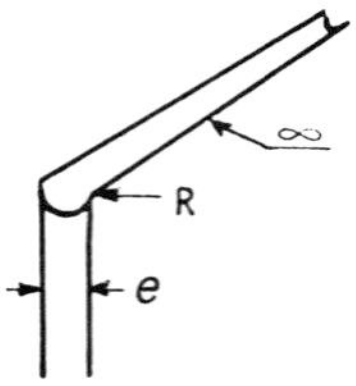

Fig. 32.126.

In the case of cracks with thickness e (Fig. 32.126), it is the width of the crack which plays the part of the pore radius. The principle radii of curvature are in one case infinite and in the other case,

$$R = \frac{e}{2} \cdot \frac{1}{\cos \theta}$$

We have:

$$P_c = T\left(\frac{1}{R} + 0\right)$$

If $\theta = 0 : P_c = \frac{2T}{e}$.

If $\theta \neq 0 : P_c = \frac{2T \cos \theta}{e}$.

32.13. Factors affecting interfacial tensions between reservoir oil and water

Temperature: interfacial tension falls as the temperature rises:

$$T \approx -aT_r + b$$

T_r = reduced temperature,
a, b = positive constants.

Pressure: interfacial tension falls as the pressure rises:

$$T = -aP + b$$

Gas dissolved in oil or water: below the bubble point interfacial tension increases with the quantity of dissolved gas.

Surface active agents: their presence in the water or oil leads to a drop in interfacial tension.

32.14. Capillary pressure in a field. Definition

Two non-mixable fluids (water, oil) are considered to be in equilibrium in a porous medium. It has been shown experimentally that for a relatively broad range of fluid saturation values each phase is continuous in the porous medium.

The porous medium is considered to be:

(a) In a state of equilibrium.
(b) In the gravitational field.
(c) At a certain temperature.

The laws of hydrostatics show that the pressure in each fluid depends solely on the elevation z:

$$\frac{dP_w}{dz} = \rho_w g \qquad \text{(Eq. 34.141)}$$

and

$$\frac{dP_o}{dz} = \rho_o g \qquad \text{(Eq. 34.142)}$$

These equations can be integrated immediately since at temperature = Const., ρ_w and ρ_o are defined functions (of P_w and P_o):

$$\frac{d\,(P_w - P_o)}{dz} = g\,(\rho_w - \rho_o) = g\Delta\rho$$

where

P_w and P_o = pressures in water and oil,
ρ_w and ρ_o = specific gravities of water and oil,
z = elevation in the field counted positively from the top downwards.

The reasoning can be extended to the entire medium which is supposed to be continuous.

Capillary pressure will be defined at every point of the porous medium by:

$$P_c = |P_w - P_o|$$

The sign of P_c is due to an arbitrary convention.

As a result of (Eq. 34.141 and 34.142) and $dP_c = g\Delta\rho\ dz$, $P_c = f(z)$:

$$P_c\,(z) = P_c\,(z_0) + g\int_{z_0}^{z} (\rho_w - \rho_o)\,dz \qquad \text{(Eq. 34.143)}$$

Capillary pressure depends upon saturation: $P_c = f\,(S_w)$.

It is necessary to examine the microscopic significance of capillary pressure.

In a state of equilibrium between the two sides of an interface separating two non-mixable fluids, there is a difference in pressure proportional to the curvature C of the interface, with the higher pressure being located on the concave side:

$$P_c = P_w - P_o = CT \qquad \text{(Eq. 34.144)}$$

$$P_c = T\left(\frac{1}{R_1} + \frac{1}{R_2}\right) = \frac{2T}{R} \qquad \text{(Eq. 34.145)}$$

T = interfacial tension characteristic of the pair of fluids,
R_1, R_2 = principal radii of curvature.

Capillary pressure is therefore due to the curvature of the interface separating the two fluids and the interfacial tension. According to (Eq. 34.143), the condition of hydrostatic equilibrium, the interface curvature is a well defined function of elevation z.

In a porous block which is sufficiently small so that on its scale it is possible to leave out of consideration the effects of gravity, there is an interface with constant curvature linked to P_c by Eq. 34.143.

According to the laws of capillarity, this interface should meet the solid surface of the porous medium at a well defined angle θ = angle of wetting.

If P_c is given, the interface between the two fluids is subject to the following conditions:

(a) Its curvature: Eq. 34.144.

(b) At the points at which it meets the solid surface the angle of junction θ is also given.

In certain simple cases (conical capillary tubes) this is sufficient to determine completely the position of the interface. The proportion of a given fluid contained in the pores will be directly linked to the capillary pressure. Arguments of this kind have led to the supposition that for a given porous medium there should be an equation of the following form:

$$P_c = f(S_w)$$

It is difficult to apply the preceding argument to a porous medium with complex geometry.

32.2. Solid wettability. Threshold displacement pressure

Let us suppose a porous medium saturated with a certain fluid into which another fluid is attempting to enter. This is the case for oil migration during geological periods or for the entry of water into an oil-bearing zone. On the other hand if the fluid in place is not the fluid of the pair which wets the solid best, the invasion of this medium by the external fluid will take place spontaneously even if it is opposed by the force of gravity.

In the following experiment (Fig. 32.21) we have a water saturated core. The walls have been made waterproof, for example by bakelization. Two ends are fitted to the core (the upper one for gas injection while the lower one is full of water and opens into a small tube at the average level of the core centre). A pipe bracket C makes it possible to determine the position of the air/water meniscus in the outlet tube.

The meniscus C remains invariable until the pressure P_d, known as threshold displacement pressure, is reached. The meniscus is displaced after P_d but a pressure P_s must be reached for the displacement to be continuous.

We have the following scheme (Fig. 32.22):

(a) The threshold displacement pressure corresponds to the inflow into the large pores.

(b) The threshold pressure corresponds to passage through the bottlenecks on the path of least resistance linking the two faces of the sample.

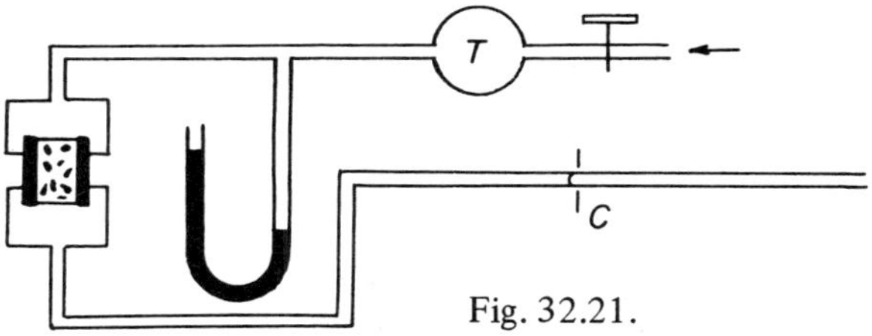

Fig. 32.21.

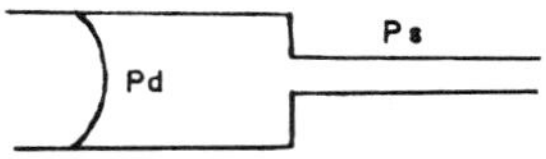

Fig. 32.22.

It can therefore be seen that, in order to make a nonwetting fluid penetrate, it is necessary to exercise a pressure above a minimum equal to the capillary pressure corresponding to the pore diameter.

According to A. Houpeurt, the following results were obtained:

Sample	k_a (mD)	ϕ(%)	Pressure (cm water) Threshold	Displacement	Radius (μ) Widening	Bottleneck	Ratio Widening Bottleneck
Sandstone (Gensac)	3.1	10.4	413	650	3.63	2.31	1.57
Coarse limestone	805	38	62	69	24.2	21.7	1.11
Folakara 2 sandstone	13.6	17.8	265	326	5.66	4.60	1.23
Folakara 2 sandstone	47.9	16.1	202	243	7.43	6.77	1.21

Figure 32.23 shows a correlation between k_a and displacement pressure P_d (water-air pair) for a group of 200 samples. (Unpublished work).

32.21. Solids and wettable reservoirs

(a) Hydrophilic: quartz, mica, carbonates, sulphates, various minerals such as sand.

(b) Hydrophobic: sulphur, talc, graphite, sulphides.

(c) Natural hydrophobic reservoirs:

Atabasca tar sands (Canada).

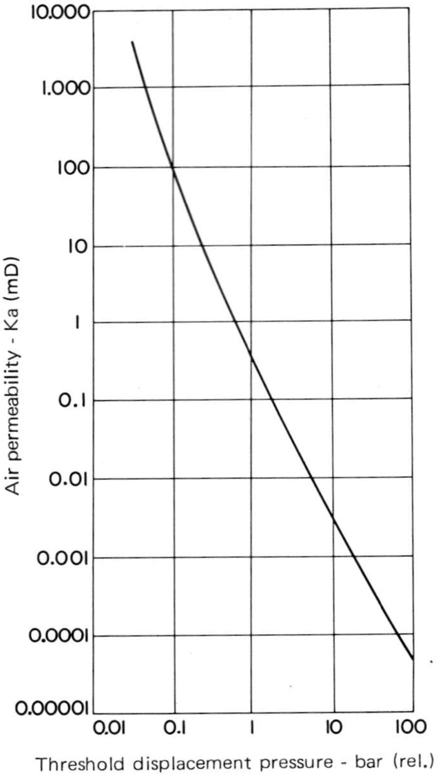

Fig. 32.23. Correlation of threshold displacement pressure and air permeability for water/air pair.
(average curve for 200 samples of different types where 3 000 > *Ka* > 0.00002 mD).

Wilcox sandstones (Oklahoma) = S_w = 0.03.

Springer sandstones (Oklahoma).

Tensly sandstones.

Remark:

Easily adsorbable polar components modify wettability and produce a wettability heterogeneity.

32.3. Morphological study of pores by mercury injection

32.31. Method and equipment (Fig. 32.311)

The equipment consists of:

(a) A mercury volume pump.

(b) A sample holder cell with a window system for observing the constant mercury level.

(c) One or several calibrated manometers.

(d) A vacuum pump and mercury manometer.

(e) A gas under pressure (for example N_2).

Operating method:

A sample for which the pore volume (and ϕ) is known is subjected to a very high vacuum (with liquid air trap) which is controlled by means of a Mac Leod gage.

The mercury level is brought to the mark (in front of the windows) with the sample completely immersed. The pump is set at zero.

The gas pressure is increased until it exceeds the mercury pressure and the latter penetrates into the sample. The constant level is reestablished in the window by means of the mercury pump. The volume of mercury injected, which is read by means of the pump, is noted for each state of equilibrium and for each pressure plateau.

Pump sampling is carried out before and after sample measurement. Pump sampling is carried out without samples, at the same pressure plateaus and at constant temperature.

The injected mercury volume is obtained for each pressure plateau by means of the differences between readings with and without samples.

Mercury saturation is calculated as a percentage of pore volume in terms of pressures in order to establish capillary pressure curves by mercury injection (Purcell method).

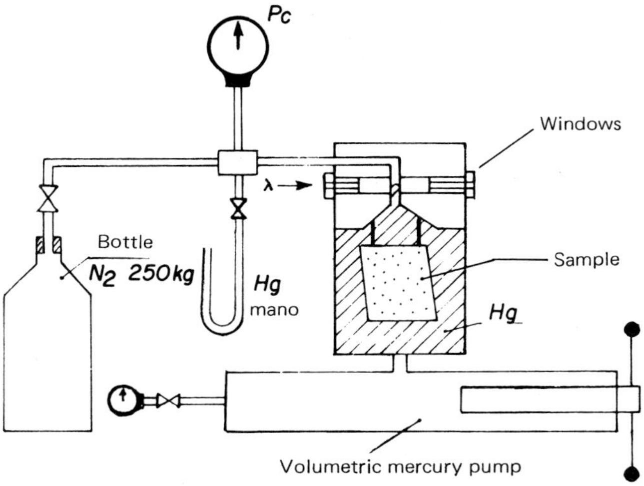

Fig. 32.311. Assembly with mercury pump for capillary pressures by *Hg* injection.

Remarks:

(a) At the beginning, at the time of the zero setting when the plug is immersed in the mercury although to a very slight degree (approximately 2 to 3 cm of mercury), the mercury may already have penetrated into the very large or macropores. This penetration can be very considerable in certain cases, amounting to as much as 30 to 40% and it must be measured.

(b) In practice, the last stage corresponds to 250 bars. At that moment most of the samples are practically 100% saturated. This pressure corresponds to pore radii of the order of 0.04 microns.

(c) The states of equilibrium can be very long in cases of low permeabilities.

32.32. Interpretation of capillary pressure curves by mercury injection

See paragr. 34.2.

32.33. Utilization of capillary pressure curves by mercury injection: concept of the most frequent pore

Let us suppose a capillary pressure curve by mercury injection: $P(s)$. The ordinate axis is divided into equal lengths ΔP corresponding to certain radii. It can be seen that the variations in saturation in these various intervals is a variable fraction of the pore volume representing a certain percentage of the latter (Fig. 32.331).

If we plot the curve of these percentages $\frac{\Delta s}{\Delta P}$ in terms of pressure P (or pore radius), a curve is obtained resembling that of Fig. 32.332. The peak of the curve corresponds to a certain pore radius called "the most frequent pore".

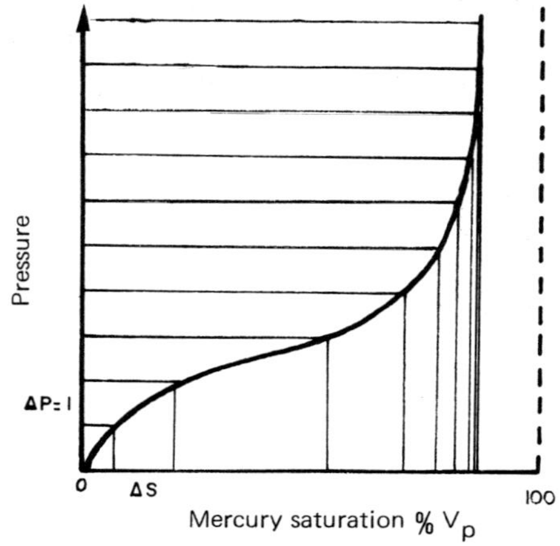

Fig. 32.331. Division of P_c curve in $\Delta P = 1$ unit.

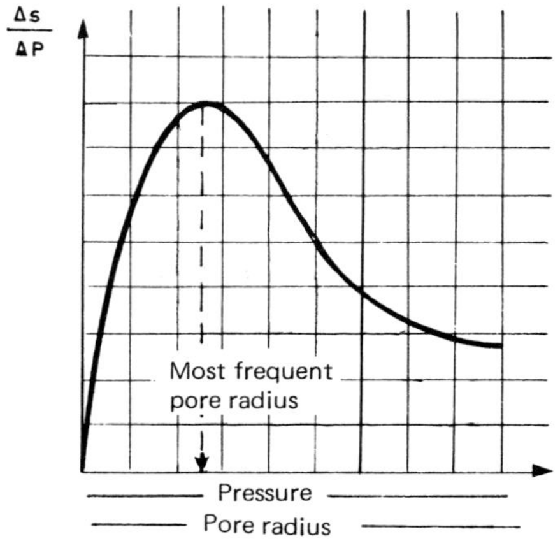

Fig. 32.332. Concept of most frequent pore radius.

The curve of accumulated frequencies can also be plotted. For each pore radius the sum of percentages corresponding to all sizes which are greater than the one under consideration are plotted. Distribution curves for pore radii are obtained.

Note:
In practical units the pore radius is given by the following equation:

$$r\,(\mu) = \frac{7.35}{P_c\ (\text{bar})}$$

$T = 480$ dyn/cm.
$\phi = 140°$.

33. OIL MIGRATION. EFFECTS OF CAPILLARY FORCES

It is admitted that the presence of hydrocarbons in a location is the result of a migration which brought these hydrocarbons from their biochemical point of origin to a place where the gravitational forces are in equilibrium with the soil reactions. This movement of a special type differs from the case in which the capillary and gravitational forces are permanently opposed in the absence of a pressure gradient in the area.

33.1. Capillary migration

Figure 33.11 represents, in a porous medium, a succession of bottlenecks and widenings with a small quantity of oil engaged in moving through a channel which was initially filled with water. This oil bubble which is lighter than water

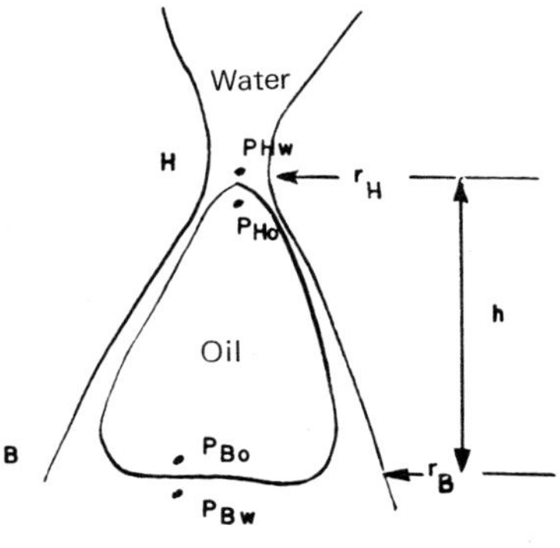

Fig. 33.11.

attempts to rise, and its face H is in front of a bottleneck with the radius r. In order to go through this bottleneck it is necessary that the oil-water contact surface take spherical form with radius r. Thus in H on each side of the meniscus there is the following difference in pressure:

$$P_{Ho} - P_{Hw} = \frac{2T}{r_H} \quad \text{(Eq. 33.11)}$$

(supposing that the rock is completely wetted by the water which is the case).

The base of the oil bubble is at B and has a shape which can be considered as spherical with the radius r_B. In the water we have:

$$P_{Bw} = P_{Hw} + h\rho_w g \quad \text{(Eq. 33.12)}$$

where

h = height of the bubble,

ρ_w = specific gravity of water,

g = gravitational acceleration.

At point B in the oil we have:

$$P_{Bo} = P_{Ho} + h\rho_o g \quad \text{(Eq. 33.13)}$$

where ρ_o is specific gravity of oil.

At point B on each side of the meniscus we have:

$$P_{Bo} - P_{Bw} = \frac{2T}{r_B} \quad \text{(Eq. 33.14)}$$

Hence the following is deduced:

$$\frac{2T}{r_B} = (P_{Ho} - P_{Hw}) - h\,(\rho_w - \rho_o)\,g$$

i.e.

$$P_{Ho} - P_{Hw} = h(\rho_w - \rho_o)\,g + \frac{2T}{r_B} \quad \text{(Eq. 33.15)}$$

If $P_{Ho} - P_{Hw} > \frac{2T}{r_H}$, the bubble will go through the bottleneck.

If $P_{Ho} - P_{Hw} < \frac{2T}{r_H}$, the bubble will not go through the bottleneck.

If we suppose $r_H = 0{,}5\ \mu$ and $T = 30$ dyn/cm, we have:

$$\frac{2T}{r_H} = \frac{2 \cdot 30}{0.5 \cdot 10^{-4}} \approx 1.2 \text{ bar}$$

which is a relatively high value.

33.2. Final field equilibrium

There are many obstacles opposing oil migration. The latter is in reality possible only as a result of the varying sizes of the bottlenecks which are presented to the bubble surface so that a lateral outlet can be found. Thus the upper part of the bubble may be blocked against the throat. The bubble will break and an isolated particle in equilibrium will be eliminated. On the other hand an additional input to this blocked bubble so that the critical height corresponding to the bottleneck is reached could make it possible for it to move again.

All these ideas can be verified in laboratory experiments with capillary channels.

In the case of a homogeneous rock, hydrocarbon saturation increases from the bottom upwards. In the case of rocks with macropores at a given elevation, it is possible that the cracks and macropores are impregnated with oil while the matrix is not. Lateral variations in facies are also found with very different impregnations: zero in the compact parts, high in the parts with larger pores. Vertically it is therefore possible to have impregnated zones separated by water-bearing layers.

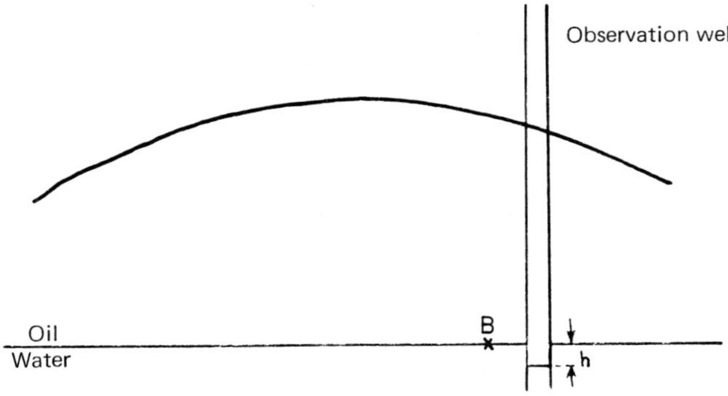

Fig. 33.21. Observation well.

If we have an observation well (Fig. 33.21) which has been completed and closed after discharge in water and hydrocarbon-bearing zones, it can be expected that fluid equilibrium will be established in this observation well because of the:

(a) Absolute pressure which should exist at the interface.

(b) Capillary pressure at the base of the accumulation.

The free level in the observation well is necessarily lower than that in the soil (the level is the same if the base of the accumulation is in a cavern). The difference in elevation h between the base of the accumulation in equilibrium and the well level is such that:

$$P_{cB} = h(\rho_w - \rho_o)g \qquad \text{(Eq. 33.21)}$$

P_{cB} = capillary pressure at the base of the accumulation zone = point B,
ρ_w = water specific gravity,
ρ_o = oil specific gravity,
g = gravitational acceleration.

The zero capillarity level takes the form of the interface in the observation well.

Capillary pressure reaches its minimum value at the base level. If this were not so the oil would have penetrated into the small pores and not into the larger pores. The result is that the base of the zone of accumulation is necessarily the site of the lowest capillary pressure which the porous medium can present, i.e. displacement pressure.

The reference elevation h is counted from the water level in the observation wells. Very often, however, there is no observation well in the field. It is possible to count h from the water/oil interface defined in another way. Capillary pressure at the interface is not zero as we have previously seen.

In general, the water/oil interface corresponds to the level at which production is 100% water. Reservoir water saturation is 100% different because of residual oil saturation.

We have seen (Eq. 34.143):

$$P_c(z) = P_c(z_o) + g\int_{z_o}^{z} (\rho_w - \rho_o)\,dz$$

i.e.

$$P_c(z) = P_c(z_o) + g(z - z_o)(\rho_w - \rho_o) \qquad \text{(Eq. 33.22)}$$

or

$$P_c(z) = P_c(z_o) + (\rho_w - \rho_o)hg \qquad \text{(Eq. 33.23)}$$

(ρ_w and ρ_o are supposed to be constant with depth if the latter varies little) and h has as its origin here the elevation z_o where $P_c = P_c(z_o)$, i.e. the level below which productions consists 100% of water. According to reservoir engineers, $P_c(z_o)$, can in practice be estimated at one of the following values:

(a) Capillary pressure where $S_o = 0$, i.e. displacement pressure.
(b) Capillary pressure at the base of the transition zone.
(c) Capillary pressure at the water/oil contact ("water level").

The value $P_c(z_o)$ are in general estimated by means of capillary pressure curves plotted in the laboratory.

33.3. Effects of a pressure gradient

If migration takes place in a rock which is the site of a flow, the migration is affected by the latter. The oil globules are subject to the buoyancy of the surrounding medium and the drive due to this medium.

It is easy to show that the accumulation of hydrocarbons does not necessarily take place at the high points of the structure and that the base of the accumulation changes from horizontal to the slope $\frac{dz}{dx}$ such that:

$$\frac{dz}{dx} = \frac{\dfrac{\partial h_w}{\partial x}}{\dfrac{\rho_w - \rho_o}{\rho_w} - \dfrac{\partial h_w}{\partial z}} \qquad \text{(Eq. 33.31)}$$

with h_w representing the hydrostatic head of the water at a reference elevation.

Oil and gas can accumulate in different places with the component due to buoyancy being very different in the two cases. The inclination of the base of accumulation is studied in further detail in basin hydrodynamics. These theoretical points of view are rarely confirmed. The underground storage at Beynes (*Gaz de France*) is an example.

In fields with thick levels and little folding, different types of accumulation can be found depending upon the intensity of the water current and oil density (Fig. 33.31).

34. INTERSTITIAL WATER DETERMINATION IN THE LABORATORY

Displacement is not complete in oil migration since there are infinitely small pores while capillary pressures are necessarily finite. It is necessary to determine the amount of water or oil which may remain in place in the field.

In order to determine water saturation in the reservoir by means of cores, the following methods can be used:

A. **Measurement of capillary pressures** by:
   (a) Restored state method.
   (b) Mercury injection.
   (c) Evaporation
   (d) Centrifuging.

B. **Measurement of water saturations** of cores with natural mud in the water. The connate water is calculated by means of empirical tables on the basis of total water saturation, gas saturation and porosity.

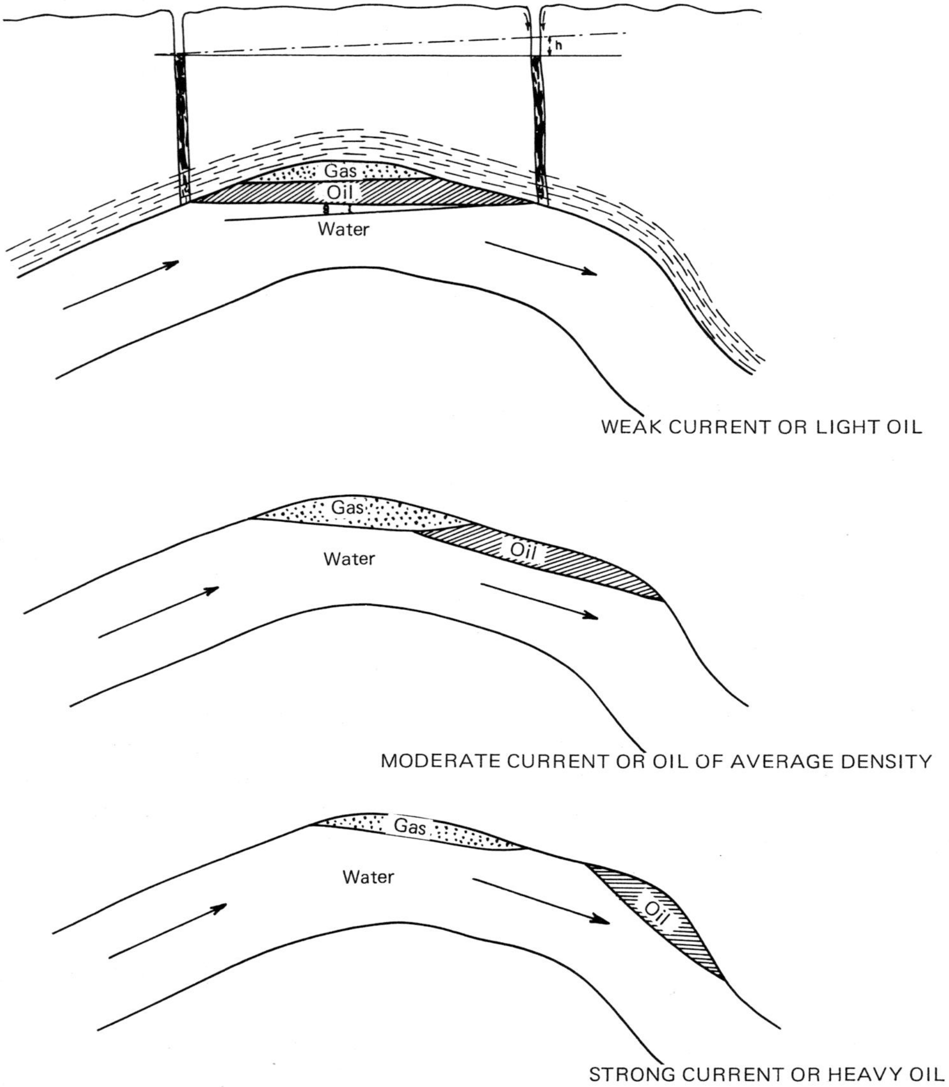

Fig. 33.31. Types of hydrodynamic accumulation of oil and gas in thick sands with little folding.
(from K. Hubbert)

C. **Measurement of fluid saturation** for cores with mud in the oil. In this case the measured water saturation represents connate water saturation. However the presence of mud makes careful verification necessary.

Measurement of capillary pressures will be considered at greater length.

34.1. Measurement of capillary pressure by the restored state method ([1])

If we suppose a rock sample from a field located at height h above the zero capillary level, the pair of fluids present in the field is characterized by:

ρ_o = specific gravity,
ρ_w = specific gravity of water,
T = interfacial tension of pair of fluids.

We have the capillary pressure:

$$P_c = h(\rho_w - \rho_o) g + \text{Const.}$$

A meniscus radius r corresponds to this capillary pressure and this pair of fluids such that:

$$P_c = \frac{2T \cos\theta}{r} = \frac{T}{r} \cdot g(S_w)$$

The relation between r and water saturation S_w is not rigorously constant if we go from one pair to another, but it will be supposed that there is an invariable relation between S_w and r. Hence the relation which is experimentally obtained in the laboratory between P'_c and S_w can therefore be validly transformed for the real pair by the equation:

$$P_c = \frac{T \cos\theta}{T' \cos\theta'} \cdot P'_c = h(\rho_w - \rho_o) g + \text{Const.} \qquad \text{(Eq. 34.11)}$$

If P'_c, T' and θ' are known for one pair, T, θ for the other pair and also ρ_w, ρ_o, the following curves can be plotted:

(a) Capillary pressure in terms of water saturation (Fig. 34.11).

(b) Or water saturation in terms of the distance h above the zero capillary level.

The height h is given by the equation:

$$h = \frac{P_c - P_c(z_o)}{(\rho_w - \rho_o) g} \qquad \text{(Eq. 34.12)}$$

([1]) Restored State Method, *Carter Oil Co., Atlantic, Texas Oil Co.,* 1947.

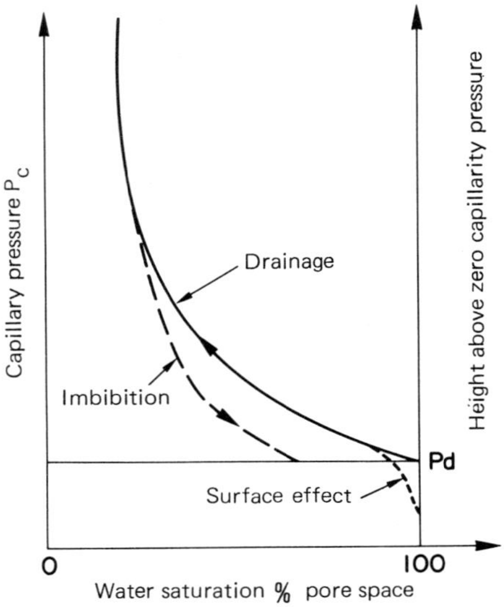

Fig. 34.11. Capillary pressure curve. (restored states method).

(a) If h is counted from the zero capillary level $= P_c(z_o) = 0$

(b) And if h is counted from the water/oil interface ("water level") previously defined (paragr. 33.2) as $P_c(z_o) \neq 0$

Remark:

Units: According to the Eq. 34.11 we have:

$$h_{(\mathrm{cm})} = \frac{P_c \quad (\mathrm{dyn/cm^2})}{(\rho_w - \rho_o)g \quad (\mathrm{cm/s^2})} \atop (\mathrm{g/cm^3})$$

$$h_{(\mathrm{m})} \approx 10.2 \, \frac{P_c \quad (\mathrm{bar})}{(\rho_w - \rho_o) \quad (\mathrm{kg/dm^3})}$$

$$h_{(\mathrm{feet})} = \frac{144\, \rho_c \quad (\mathrm{psi})}{(\rho_w - \rho_o) \quad (\mathrm{lb/ft^3})}$$

When a sample is drained by progressively increasing capillary pressure by stages, it can be observed that gas permeability remains zero until a certain pressure is reached (displacement pressure). It builds up then as capillary pressure rises (Fig. 34.11).

During imbibition, gas permeability falls in a different way than during draining and becomes zero for a capillary pressure value which is not zero and has always been found to be in the neighborhood of displacement pressure. It is probable that below this capillary pressure the nonwetting phase is not continuous and thus cannot be attributed a pressure value. Capillary pressure is no longer defined.

On the capillary pressure curve the following can be distinguished (Fig. 34.12):

(a) Saturation near 100% corresponding to the water level (measurement takes place at the sample surface).

(b) Zone of transition.

(c) Region of irreducible saturation corresponding to vertical part. The irreducible water saturation is affected by clayey particles: there is water fixation which is not determined by the laws of capillarity.

Capillary pressure P_c depends upon:

(a) Interfacial tension $= T$.
(b) Angle of contact $= \theta$.
(c) Rock permeability $= k$.
(d) Rock porosity $= \phi$.

The curves in Figs. 34.13, 34.14, 34.15, 34.16 represent variations of P_c with T, θ, k, ϕ.

The curves in Fig. 34.17 show interstitial water distributions in terms of permeability for various porous media.

The assembly utilized for the restored state method in order to displace the wetting stage is shown in Fig. 34.18.

We have:

(a) A steel cell for imposing pressure.

(b) A semi-permeable wall C (principal part of the assembly) which is a fine porcelain wall whose large pores are small enough so that they are permeable only at pressures above 5 bars.

(c) Manometer and source of gas saturated with water vapor.

The brine saturated wall is in contact with a chamber containing brine. The water phase is entirely at atmospheric pressure. The samples which are 100% saturated in this brine are placed on the wall with the sample-wall contact being as complete as possible (very fine diatom powder). Above the wall, the cell is filled with air at constant pressure. At each pressure plateau (of the order of 48 h) the following are noted:

(a) The brine volume produced (capillary tube).

(b) Or the variation in weight.

It is then possible to determine the relation P_c in terms of brine saturation (S_w expressed as percentage of pore volume). The pressure plateaus which are frequently used are : 0 - 0.03 - 0.07 - 0.14 - 0.28 - 0.56 - 1 - 2 - 3 - 4 bars rel. (in the case of the water/air pair capillary pressure of 3 bars corresponds to a

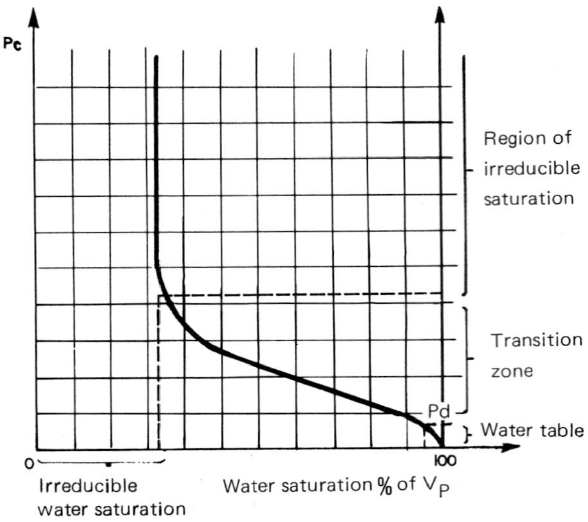

Fig. 34.12. Typical capillary pressure curve.

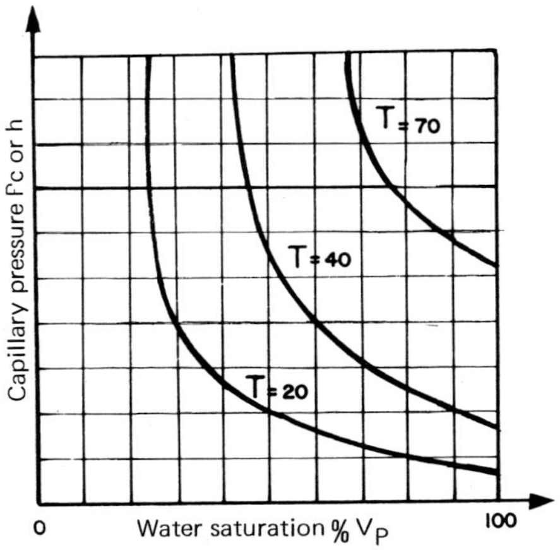

Fig. 34.13. P_c versus T.

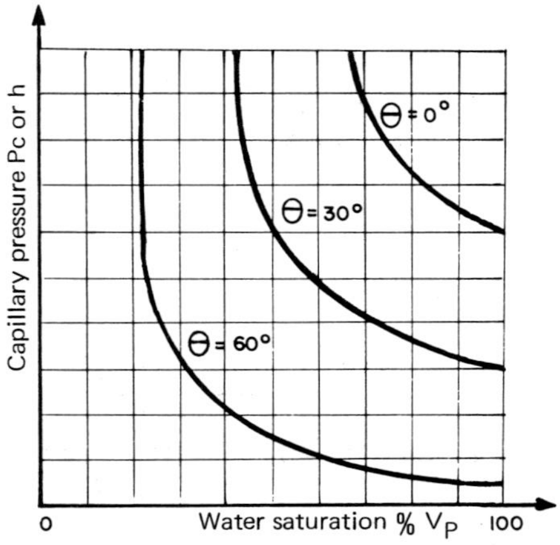

Fig. 34.14. P_c versus θ.

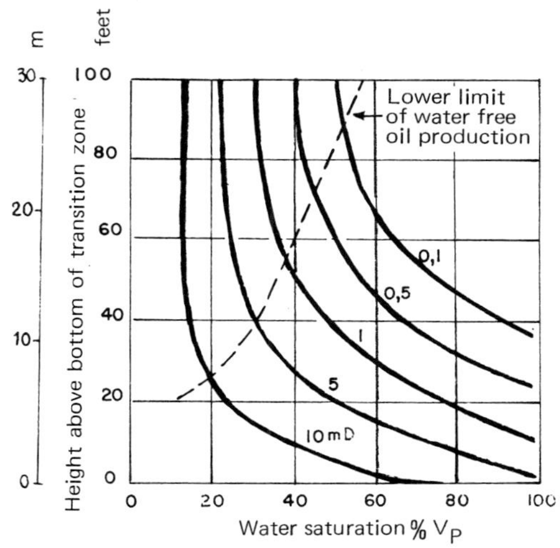

Fig. 34.15. Capillary pressure versus k_a.

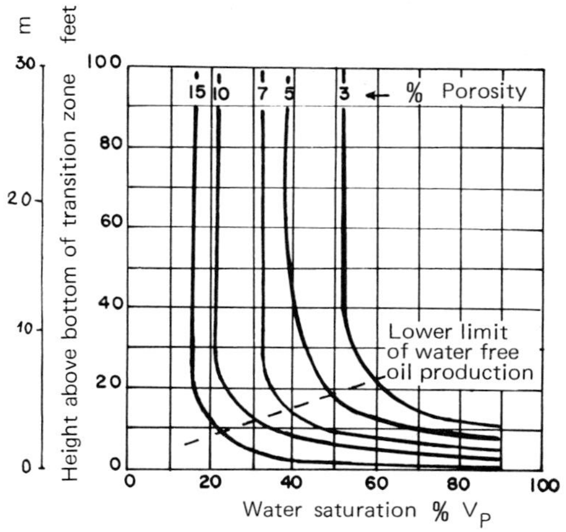

Fig. 34.16. Water distribution curves.

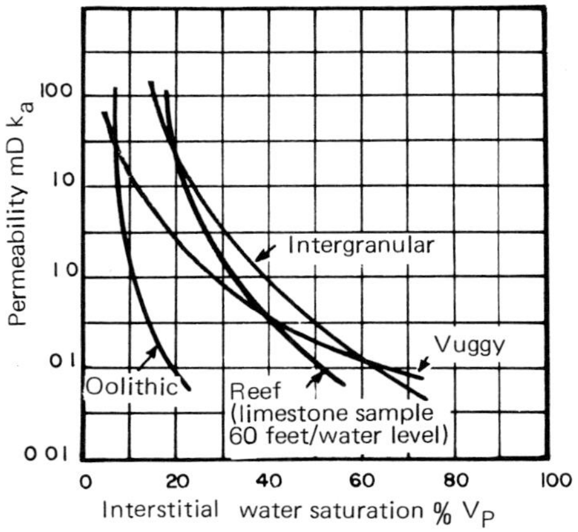

Fig. 34.17. Typical water distribution.

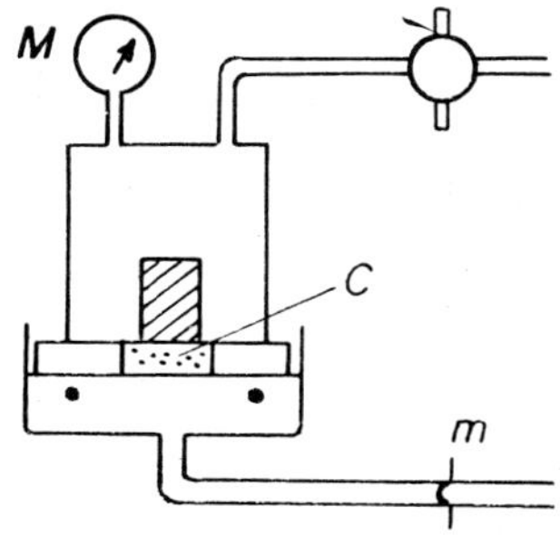

Fig. 34.18. Assembly with porous plate for restored states.

height above the zero capillary level of the order of 30.5 m for the water/oil pair).

Although the determination of capillary pressures by the restored state method may be slow and complex, it is an excellent method. It is highly recommended especially if the samples are slightly clayey.

34.2. Measurement of capillary pressure by mercury injection (Purcell method) ([1])

We have already described (paragr. 32.3) the apparatus and method for determining capillary pressures by mercury injection.

In the case of the water/air pair where the wetting phase is displaced, there is always a path through which circulation is possible leading to the creation of irreducible saturation in the pores which do not participate in this circulation.

In the case of the mercury/air pair, this condition does not exist and the curves obtained do not show the characteristic asymptote for irriducible saturation. The latter can perhaps be defined by the saturation corresponding to the beginning of the rectilinear part of the curve (if this point is marked) or, if not, to the saturation corresponding to a pressure of, for example, 10 to 15 or 20 bars depending upon the samples.

The shape of the curves obtained by plotting the equation for capillary pressure as a function of mercury saturation expressed as a percentage of pore volume is very variable from one sample to another (pore volume is carefully determined by an appropriate method such as immersion in a solvent).

Figure 34.21 concerns a homogeneous matrix medium. The beginning of the curve corresponds to a surface effect, i.e. the mercury has not yet definitely entered the pores. By means of a "surface correction" it is clearly possible to eliminate this part. The part represented by a broken line is then obtained.

In certain cases where the pore radii are small, the Ps_1 threshold can be high: up to 30 bars abs., and even more.

Figure 34.22 makes it possible to distinguish:

(a) The macropores, the part $O - a$ which are invaded under very low pressures.

(b) The pores constituting the matrix (as above) which are relatively regularly distributed. The part *ab* corresponds to channels which can be used for circulation while the part *bd* corresponds to the windings of the channels.

Figure 34.23 corresponds to the case of two homogeneous matrix media separated by an intermediate medium, for example calcite, coating the large

([1]) *Shell Oil Co.*, 1940

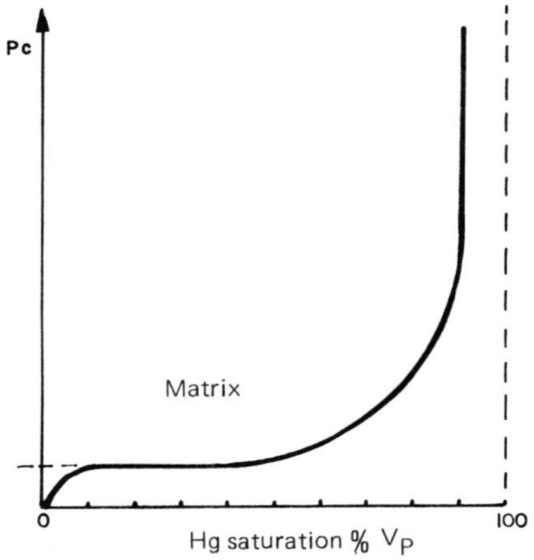

Fig. 34.21. Capillary pressure by mercury injection: homogeneous matrix medium.

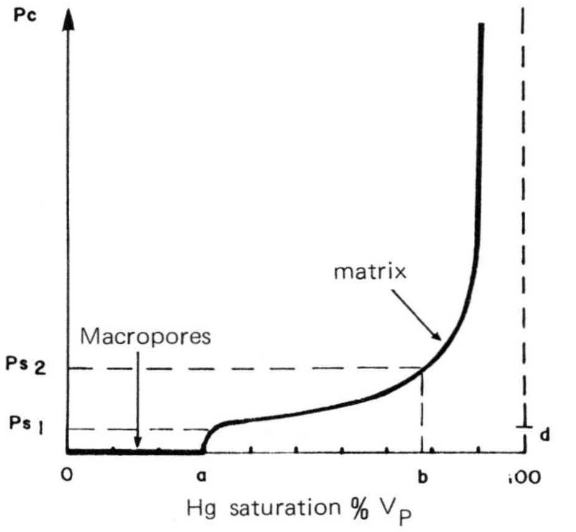

Fig. 34.22. Capillary pressure by mercury injection: medium with macropores and matrix.

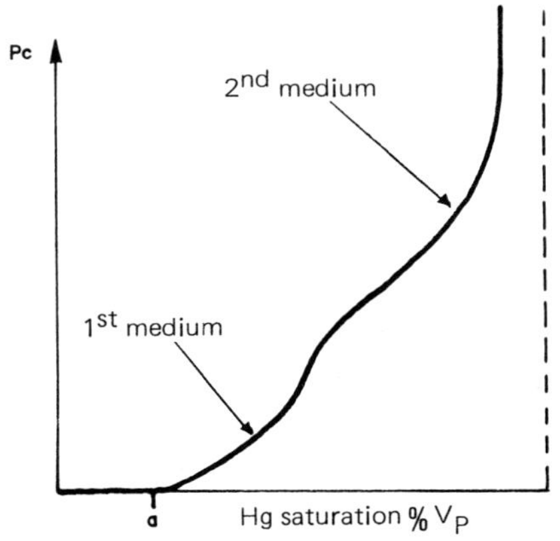

Fig. 34.23. Capillary pressure by mercury injection in the case of 2 matrix.

pores. The tangential departures of the curve from the abscissa should also be noted. This shape can be interpreted by observing that the macropores have later been filled by another medium.

The speed and accuracy of this method account for the fact that it is very widely used. It is however not suggested for very clayey samples since for a given pressure the saturation is too high because the pores are invaded by clay.

34.3. Measurement of capillary pressure by evaporation (Messer method) (1)

This method is utilized principally in the laboratory in order to determine irreducible saturation. The sample is saturated in toluene or water, etc. and is subjected to a current of dry air which may or may not be hot. The weight losses of the sample in a box hanging from a balance are noted in terms of time. The weight in terms of the time curve shows an angular end point (Fig. 34.31). The weight corresponding to this angular point is used for calculating the irreducible water.

At the beginning, surface evaporation is fed by the interstitial water and there is a given slope. When the irreducible water is reached there is evaporation only by diffusion so that the slope is much flatter (method used by *Continental Oil Co., Bureau of Mines, Elf,* etc.).

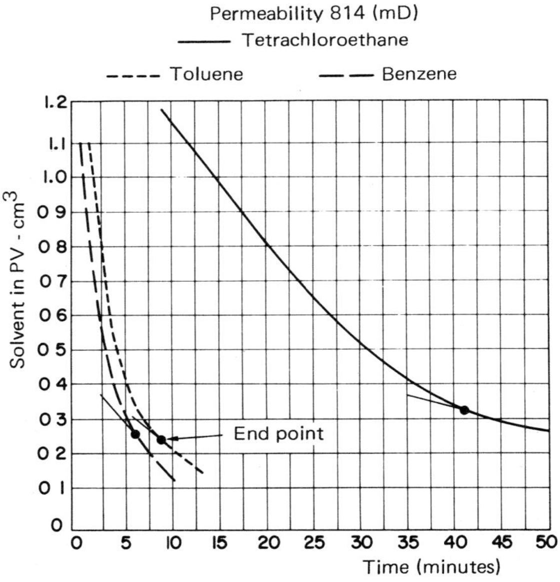

Fig. 34.31. Evaporation curve. (from P. Albert).

(1) *Continental Oil Co.*, 1951.

34.4. Measurement of capillary pressure by centrifuging ([1])

The brine saturated sample with density ρ_1 is placed in a centrifuge rotating at n RPM. The sample is located within a medium with the specific gravity ρ_2 (air or oil). The maximum pressure difference on the end face of the core is given by the equation:

$$\Delta P = A \cdot (\rho_1 - \rho_2) n^2 h \cdot \gamma \qquad \text{(Eq. 34.41)}$$

where

ΔP = the pressure difference on the end face and

$P_m = \dfrac{\Delta P}{2}$,

h = centrifuge plate radius,

γ = constant depending upon instrument,

A = constant depending upon units.

In the case of small samples with n = 16 000 RPM and $\rho_1 - \rho_2 = 1.064 - 0.754 = 0.310$, we have found for a given centrifuge that $\Delta P = 11.95$ bars and

$$P_m = \frac{\Delta P}{2} \approx 6 \text{ bars.}$$

Case of application

Certain samples have swelling marls and very low k_a. In order to have a point well above the maximum possible pressure with semi-permeable walls (4 to 5 bars), the centrifuging method is used which makes it possible to show one or two points as high as 21 bars.

34.5. Comparison of different methods for determining interstitial water by capillary pressure measurements

(a) **Restored State**

The method gives the entire curve $P_c = f(S_w)$. Irreducible water saturation is determined on the curve. If the samples are relatively clayey the method is preferable to others.

The disadvantage lies in its length of time from 3 weeks to 1 month, for the description of a complete curve (routine measurement). It is necessary to have batteries of instruments or semi-permeable walls which can hold several samples.

(b) **Mercury injection**

This is a rapid and precise method and is very useful for the analysis of pore morphology. It does not give directly the irreducible water saturation. It is not recommended in the case of clayey samples.

([1]) *Atlantic Refining Co.*, 1951.

(c) **Evaporation**

This method is used only for determining rapidly the irreducible water saturation.

(d) **Centrifuging**

This method is used only for obtaining a point corresponding to a high pressure.

34.6. Conversion of laboratory results

Capillary pressure measurements made in the laboratory with mercury/air or water/air pairs must be converted for the water/oil pair which exists in the field.

If we suppose that the average curvature of an interface in a rock is solely a function of fluid saturation, the ratio of capillary pressures in the case of the mercury/air and water/air pairs is:

$$\frac{P_c\ H_g/\text{air}}{P_c\ \text{water/air}} = \frac{480}{70} = 6.57 \qquad \text{(Eq. 34.61)}$$

Experience has shown that we have:

(a) For limestones: a ratio of the order of 5.8.
(b) For sandstones: a ratio of the order of 7.5.

There is no common factor for all rocks.

In order to use the laboratory results for capillary pressures, it is necessary to convert them to reservoir conditions. The laboratory results are obtained with a gas/water, oil/water system which should have the same physical properties as the water, oil or gas of the reservoir. Two techniques differing only by their initial hypothesis are used for converting the laboratory capillary pressure results for reservoir conditions:

$$(P_c)_{\text{reservoir}} = \frac{T_{\text{water/oil}} \cos\theta_{\text{water/oil}}}{T'_{\text{water/gas}} \cos\theta_{\text{water/gas}}} \cdot (P'_c)_{\text{laboratory}} \qquad \text{(Eq. 34.62)}$$

or

$$(P_c)_{\text{reservoir}} = \frac{T_{\text{res.}}}{T'_{\text{lab.}}} (P'_c)_{\text{laboratory}} \qquad \text{(Eq. 34.63)}$$

Examples

1. Conversion of water/air pair (laboratory) to water/oil couple (reservoir). We have:

In the laboratory	In the reservoir
Capillary pressure $= P'_c$	P_c
$T'_{water/air} \approx 70$ dyn/cm	$T_{water/oil} \approx 28$ dyn/cm
$\theta'_{water/air} \approx 0°$	$\theta_{water/oil} \simeq 33$ to $55°$
	$\cos\theta = 0.869$ to 0.643

hence

$$(P_c)_{\substack{water/oil\\ reservoir}} = \frac{28 \times 0.869}{70} \times (P'_c)_{\substack{water/air\\ laboratory}} \qquad \text{(Eq. 34.64)}$$

i.e.

$$\frac{P_c}{P'_c} \approx \frac{1}{3 \text{ to } 4}$$

2. Conversion of mercury/air pair (laboratory) to water/oil pair (reservoir). We have:

In the laboratory	In the reservoir
P'_c	P_c
$T'_{mercury/air} \approx 480$ dyn/cm	$T_{water/oil} \approx 28$ dyn/cm
$\theta'_{mercury/air} \approx 140°$	$\theta_{water/oil} \approx 33$ to $55°$

hence

$$(P_c)_{\substack{water/oil\\ reservoir}} = (0.066 \text{ to } 0.05)\,(P'_c)_{\substack{mercury/air\\ laboratory}}$$

i.e.

$$\frac{P_c}{P'_c} \approx \frac{1}{15 \text{ to } 20}$$

In reality the conversion factor depends upon saturation.

34.7. Average results for capillary pressures and correlations

Two methods are used for establishing a correlation between capillary pressure results for the same geological formation.

34.71. The function of capillary pressure

Some authors have attempted to link the permeability of rocks to their morphology. Within a more limited field, Leverett believes that different samples from the same sediment have linked properties so that there exists an invariant specific to the sediment. He has introduced the function $J(S_w)$ known as the capillary pressure function which is a dimensionless grouping of the physical properties of the rock and the fluids saturating it:

$$J(S_w) = \frac{P_c}{T} \cdot \left(\frac{k}{\phi}\right)^{1/2} \qquad \text{(Eq. 34.711)}$$

Some authors write it:

$$J(S_w) = \frac{P_c}{T\cos\theta} \times \left(\frac{k}{\phi}\right)^{1/2} \qquad \text{(Eq. 34.712)}$$

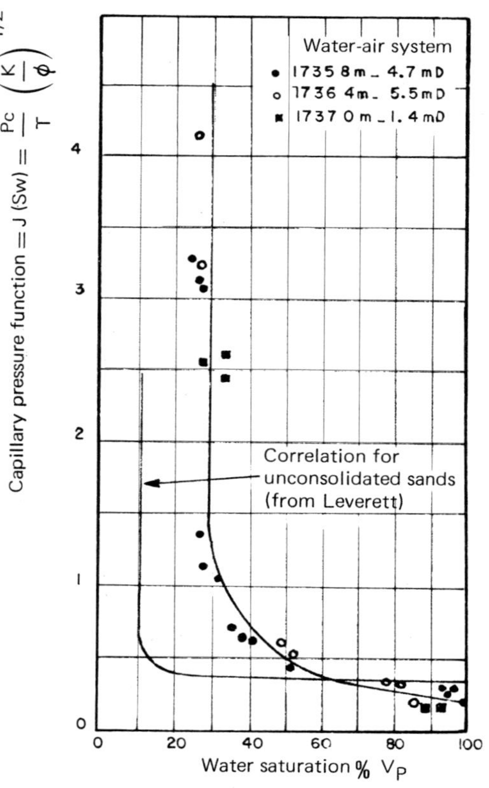

Fig. 34.711. Capillary pressure function in Weber consolidated sandstone (from W. Rose and W.A. Bruce).

It should be noted that this function $J(S_w)$ is proportional to a quantity which is in relatively direct relation to the pore radius. Function J was at first proposed as a means of conversion for all the capillary pressure results to a general curve. There are considerable differences in the correlation between function J and water saturation from one formation to another, which makes it impossible to plot a general curve, but for the same formation a very good curve can be plotted.

The example shown in Fig. 34.711 (according to W. Rose and W.A. Bruce) is the function $J(S_w)$ plotted on the basis of capillary pressure curves for various samples in the case of the water/air pair.

34.72. Second correlation method for calculating connate water saturation

Capillary pressure measurements are made for a selection of representative samples. The selection is made in terms of permeability (Fig. 34.721) according to Wright and Wooddy. As a first approximation, for the establishment of capillary pressure correlation, the water saturation curves as a function of log k_a are plotted for constant capillary pressure values (arbitrarily chosen values). For each capillary pressure value a curve of the following form is thus obtained:

$$S_w = a \log k + \text{Const.} \qquad \text{(Eq. 34.721)}$$

If the average value of k in the reservoir is known it is then possible to determine the average capillary pressure curve according to the results of permeability distribution for the field.

The capillary pressure curves (Fig. 34.721) give the correlation $S_w = a \log k + \text{Const.}$ (Fig. 34.722). The tendancy of all capillary pressure curves to converge for high permeability values should be noted.

34.73. Correlation of irreducible water saturation and permeability-porosity

We have seen in paragraph 34.1, Figs. 34.16 and 34.721, that capillary pressures depend upon physical characteristics of the reservoir rock. For field analysis purposes it is possible to indicate, for a given formation, a correlation between irreducible water saturation: S_{iw} (vertical part of capillary pressure curves by restored state method), permeability k and porosity ϕ. In general a correspondence of the following form is used (Fig. 34.731):

$$S_{iw} = a \log k + C \qquad \text{(Eq. 34.731)}$$

$$S_{iw} = a_1 \phi + a_2 \log k + C \qquad \text{(Eq. 34.732)}$$

$$S_{iw} = a_1 \phi + a_2 \phi^2 + a_3 \log k + a_4 (\log k)^2 + C \qquad \text{(Eq. 34.733)}$$

$a \ldots C$ are constants.

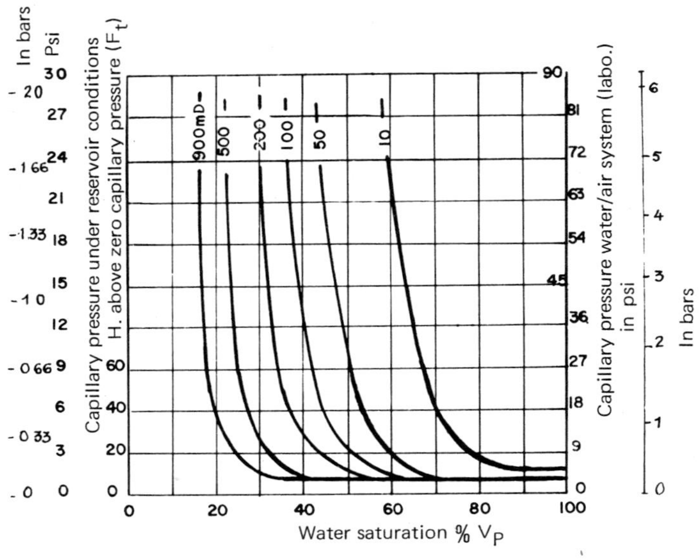

Fig. 34.721. Reservoir fluid distribution curves.
(from Wrigth and Wooddy).

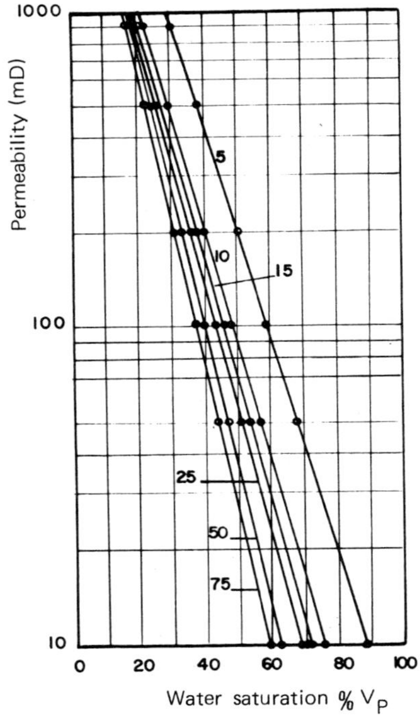

Fig. 34.722. Correlations of water saturations and permeabilities for different capillary pressures.
(from Wright and Wooddy).

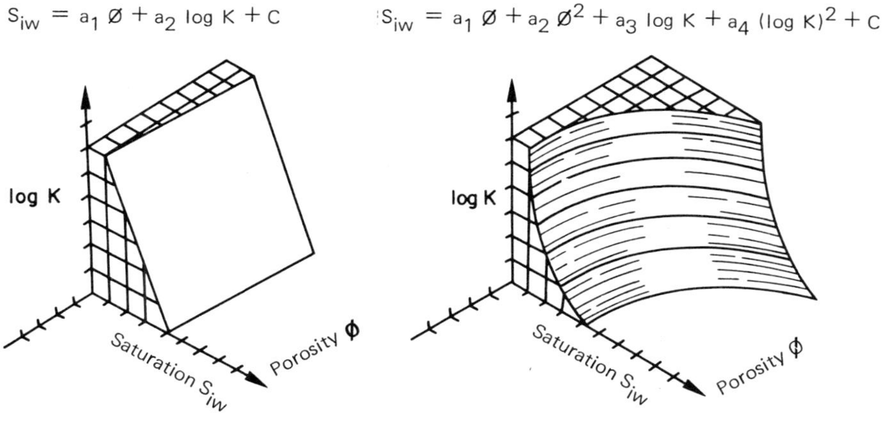

Fig. 34.731. Correlation of S_{iw} and ϕ, K.

35. IN SITU DETERMINATION OF INTERSTITIAL WATER SATURATION (WITH ELECTRIC LOGS)

35.1. Basic situation

We have seen (Chapter 1) that the formation factor $FF = \frac{R_o}{R_w}$ (R_o resistivity of sample 100% saturated with brine whose own resistivity is R_w) is linked to porosity ϕ by an equation of the form:

$$FF = \frac{a}{\phi^m}$$

where a and m are constants characterizing the rock (m varying from 1.3 to 2.2 and more, depending upon the state of cementation of the reservoir).

Since oil is an electrical insulator, it can be seen that the fact that a certain quantity of water is replaced by oil in the rock means an increase in resistivity.

Archie has shown experimentally that between the true resistivity (R_t) of the rock partially saturated with oil, the value S of the water saturation corresponding to this resistivity and the resistivity R_o of the rock 100 % saturated with oil there is the following equation:

$$S^n = \frac{R_o}{R_t} = RR \text{ (resistivity ratio)} \qquad \text{(Eq. 35.11)}$$

which can be written:

$$S^n = \frac{(FF)\, R_w}{R_t}$$

$n \approx 2$, if the rock is water wet,
$2 < n < 4$, if the rock is oil wet.

This equation makes it possible to obtain the *in situ* rock interstitial water saturation on the basis of resistivity measurements.

35.2. Case of a water-bearing zone

If the formation is homogeneous and is visibly oil-bearing at the top and water-bearing at the base electrical logs make it possible to determine R_o and R_t immediately.

35.3. Determination by means of Formation Factor

(a) The logs give R_t.
(b) Micrologs or microlaterologs give *FF*.
(c) R_w is determined directly or by means of the *SP*.
(d) Hence $R_o = (FF) \cdot R_w$ and $S^2 = \frac{R_o}{R_t}$.

If (*FF*) is not known it can be replaced by porosity data (measured on the core or according to the neutron curve).

35.4. Tixier method

This method based upon the existence of the invaded zone is used for relatively compact formations. It utilizes the hypothesis that there is a relation between residual oil saturation in the zone invaded by mud filtrate and the original saturation. An equation of the following form is supposed:

$$S_i^2 = S \qquad \text{(Eq. 35.41)}$$

where S_i is the saturation of the zone invaded by conductor fluids. If R_z is the resistivity of these conductor fluids (filtrate + field water), Archie's formula in modified form can be applied to the invaded zone:

$$S_i^2 = \frac{(FF) \cdot R_z}{R_i} \qquad \text{(Eq. 35.42)}$$

Hence

$$S = \frac{R_i}{R_t} \cdot \frac{R_w}{R_z} \qquad \text{(Eq. 35.43)}$$

R_t is given by a resistivity curve with wide spacing,
R_i is given by a resistivity curve with narrow spacing,
R_z is determined by direct measurements on the filtrate and the field water.

The Spontaneous Potential (SP) is linked to the resistivity of the field water and there is a relation between $\frac{R_w}{R_z}$ and the SP value. This is resolved by means of charts where the SP value and the ratio $\frac{R_i}{R_t}$ are entered directly (see Collection of *SPE-Schlumberger,* etc. charts).

35.5. Rapid simplified approach

The utilization of Archie's formula requires relatively complex calculation, excluding all continuous interpretation which is not done by computer. A rapid simplified approach for continous quantitative analysis of water saturation by means of logs has been presented by J. Raiga-Clémenceau.

Returning to the idea of the ratio of water volume to pore volume facilitates the approach by breaking down the parameters being treated. It should be noted that the water porosity of a reservoir $\phi_w = \phi \cdot S_w$ can be read directly on a real resistivity log simply be applying the appropriate calibration. The real resistivity curve can be plotted in terms of porosity on a total porosity log by means of a change in scale which is shown graphically.

A resistivity log can be considered as a water porosity log ϕ_w. We have:

(a) Formation factor relative to the water fraction of total porosity alone:

$$F_w = \frac{R_t}{R_w} = \frac{a}{\phi_w^m} \approx \frac{1}{\phi_w^2}$$

(b) ϕ_w is deduced from an appropriate chart of the *Schlumberger* Collection (C 10):

$$\phi_w \approx \sqrt{\frac{R_w}{R_t}}$$

The final objective is to find oil porosity (and oil saturation) : $\phi_o = \phi - \phi_w$.

The speed and simplicity of the operation make it possible to use the results in the place where the recordings were made within thoroughly satisfactory limits of accuracy.

36. INDIRECT METHODS FOR DETERMINING FLUID SATURATION BASED UPON ROCK ANALYSIS

There is a tendancy to confuse the interstitial water content of an *in situ* core with the water content it has when it arrives at the laboratory after removal from the core barrel. The two are not related.

The saturation values are modified by the following causes:

Invasion by mud filtrate is shown *in situ* by a reduction in oil saturation and an increase in water saturation.

While the core is being brought from the bottom to the surface, the fall in pressure results in a fall in oil pressure and the appearance of a gas phase as a result of the expansion of the dissolved gases. There is then a fall in oil saturation (shrinkage) and a reduction in water saturation (expansion of dissolved gases and expulsion).

Evaporation is inevitable during handling or transport if the necessary precautions were not taken.

The two tables below schematically show these invasions in the cases of two different types of mud: water base mud and oil base mud.

(a) Variation in fluid saturation for a core between the reservoir and the surface in the case of **water base mud:**

Saturation	Oil	Gas	Water
At surface	12% ↑	40% ↑	48% ↑
	shrink	expand	expulse
In core barrel	15%	0	85% ↑
	flush ↑		invade
In reservoir	70%	0	30%

(b) Variation in fluid saturation of a core between reservoir and surface in the case of **oil base mud:**

Saturation	Oil	Gas	Water %
At surface	↑ 40%	↑ 30%	30
	shrink and expulse		
In core barrel	70%	0	30
	↑ invade		
In reservoir	70%	0	30

It is clear that the measurement of fluid saturation can be carried out only on fresh or very well protected samples. In no case should the samples be washed when removed from the core holder or exposed for long.

The fluids contained in the core whose saturation is to be measured are therefore:

Gas.
Oil.
Water.

36.1. Determination of gas saturation

This determination is done by injecting mercury into a fresh core of known volume. The procedure will be described below under the so-called fluid summation method. The equipment used is a mercury volume pump.

36.2. Determination of water and oil saturation by distillation at atmospheric pressure (Retort method of fluid summation, Core Laboratories process, Patent Nos. 2 282 654 and 2 361 844)

A known weight and volume of fresh samples from the core center is coarsely ground and placed in a hot oven (Fig. 36.21) whose temperature is controlled. Oil and water are collected in test tubes by distillation.

If the rock volume and the collected water and oil volumes are known, it is possible to determine the values of the fluids as a percentage of rock volume. Saturation is calculated by means of porosity.

In this method it is necessary to note:

(a) High temperatures produce dehydration of clays. The quantity of water collected will be too great. It is therefore necessary to establish in advance a volume of water collected-heating time curve (Fig. 36.22) in order to avoid this disadvantage by determining the plateau above which the clays are dehydrated.

(b) Part of the oil is lost by cracking and, it is also necessary to establish a measured oil-real oil correction curve depending upon the oil found (or if not, an average curve) (Fig. 36.23).

In spite of the need for these sampling curves and the possibility of emulsions, this method has the following advantages:

(a) Saturation is determined for large representative fragments selected from the center of the fresh core.

(b) The volume of each fluid collected is measured directly and not estimated by differences.

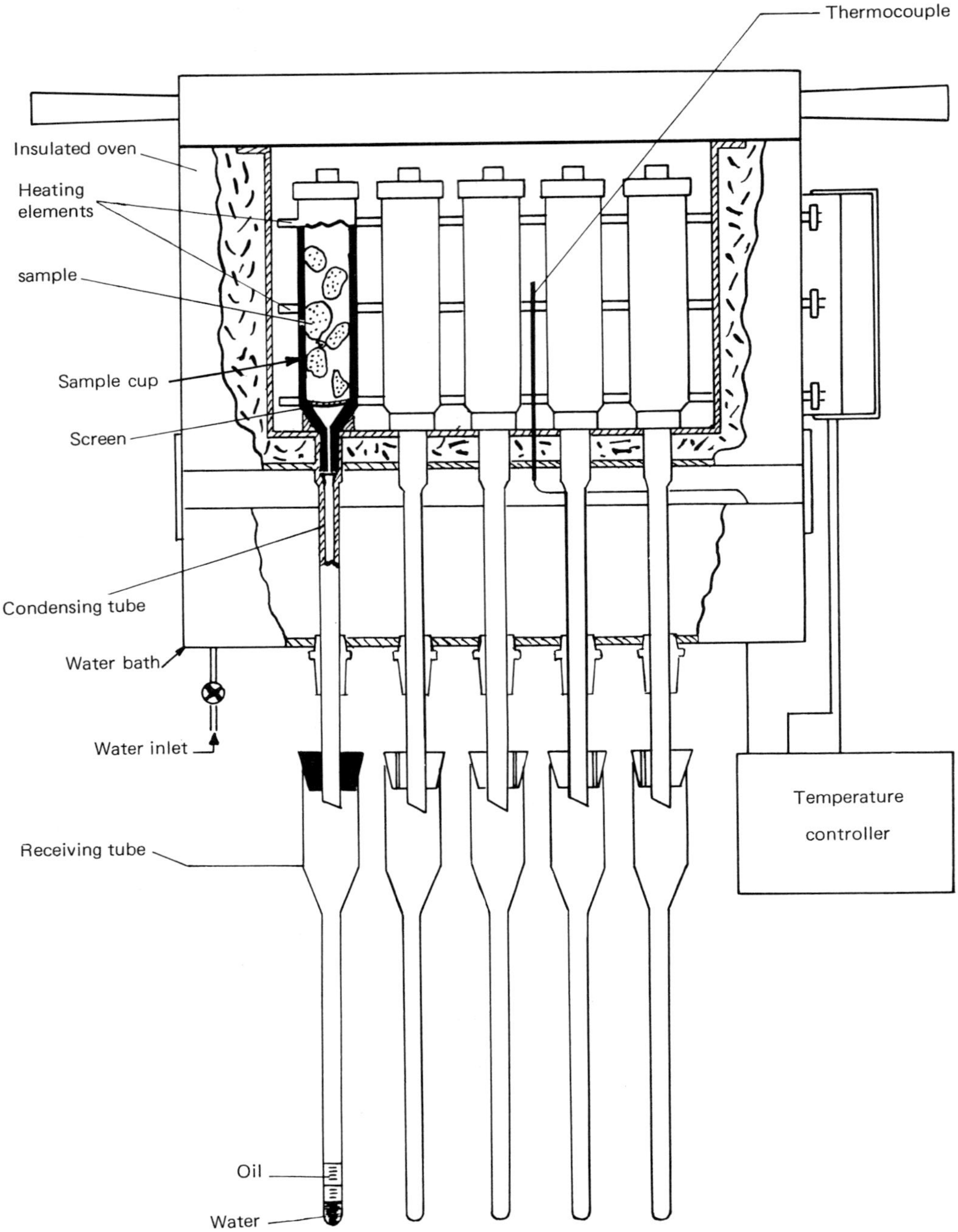

Fig. 36.21. Oven retort atmospheric pressure. (*Core Laboratories Inc.* type retorts).

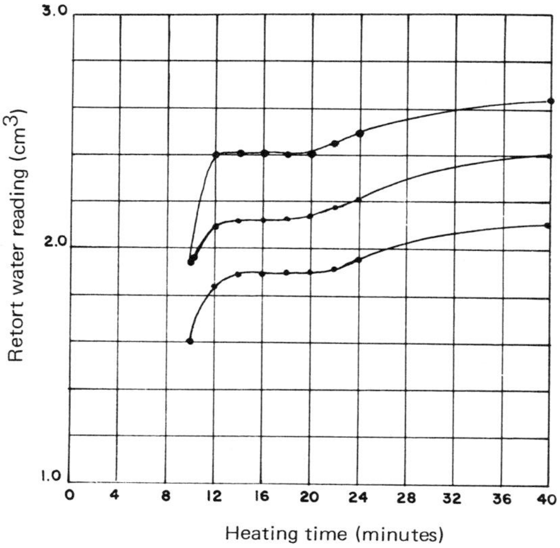

Fig. 36.22. Water calibration curves.

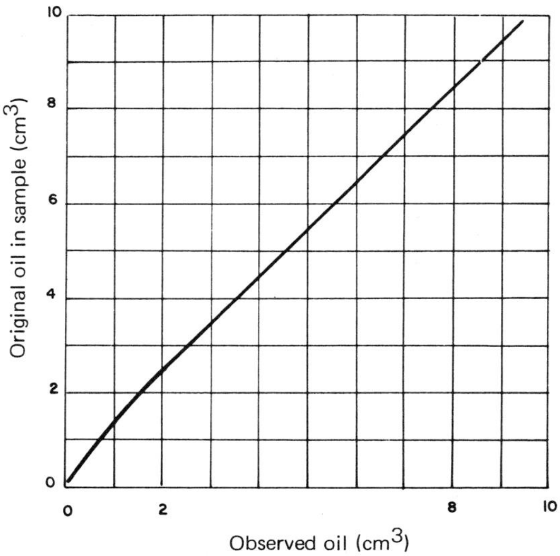

Fig. 36.23. Oil calibration curve.

(c) There is no error because of salt deposits.

(d) There is no loss of material in handling.

(e) It is a very rapid method which gives results having a high degree of precision.

36.3. Determination of water and oil saturation by vacuum distillation

The fresh sample containing its fluids is vacuum distilled. The method used for the analysis of whole or large cores will be described below. Figure 36.31 shows a diagram of the assembly. The fluids collected after condensation (mixture of methyl alcohol and dry ice): water + oil are subject to the same corrections as above.

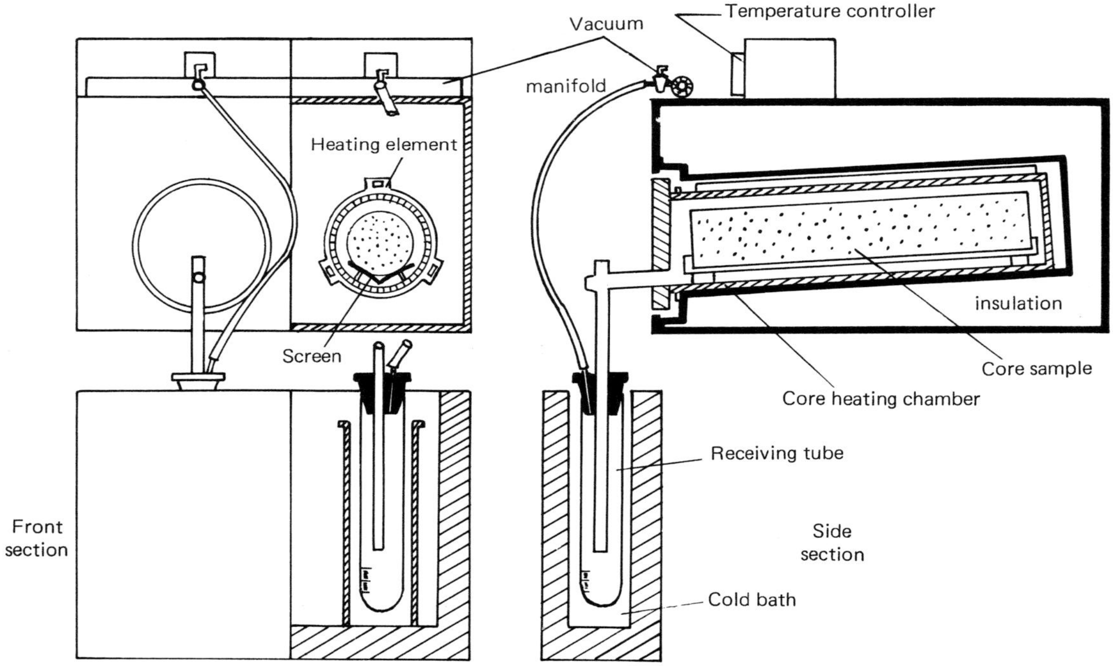

Fig. 36.31. Vacuum retort.

36.4. Determination of water content by solvent extraction (Dean-Stark distillation)

The Bureau of Mines has developed a process which combines extraction by diffusion with distillation. The apparatus (Fig. 36.41) consists of:

(a) A flask in which the core is placed.

(b) An ASTM trap in which the water is collected which is poured off after condensation of the distilled azeotropic mixture.

The problem consists in the collection in the trap of water which should be the only substance to be found there. There are many sources of error and the method is not always very effective, while extraction time is very long (several days to several weeks). The quantity of water is read directly.

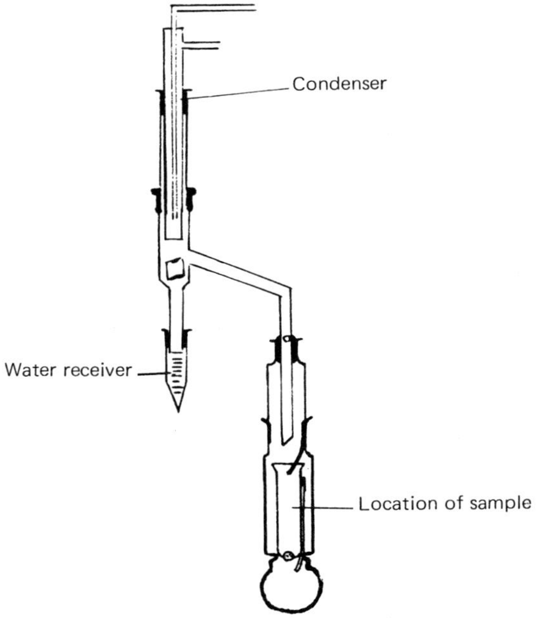

Fig. 36.41. ASTM. Distillation extraction apparatus (modified by RAP).

36.5. Determination of water content by reaction with calcium hydride

This method was developed and used principally by S.N. Repal and SNPA. The sample is placed in a test tube with inert material. 3 g of calcium hydride in powder form are placed in the tube and the tube is connected to an assembly for measuring the volume of gas discharged. The tube is immersed in an oil bath which brings the sample temperature from 40° to 135° C in approximately one half hour.

If the sample is clayey, the final temperature will be < 110° C.

Heating time is 2 to 3 hours for average permeabilities, 8 hours for very low permeabilities.

The following very exothermic chemical reaction occurs:

$$CaH_2 + 2H_2O \rightarrow Ca(OH)_2 + 2H_2$$

If the temperature is high we then have:

$$CaH_2 + H_2O \rightarrow CaO + 2H_2$$

The measured volume of H_2 (corrected for temperature and atmospheric pressure) is much higher in the second case for the same quantity of water.

36.6. Determination of oil content by colorimetry

(a) Firstly a colorimetric scale is established with test tubes containing a solvent and increasing quantities of crude following a carefully selected progression.

(b) A known quantity of rock is ground and placed in a test tube containing solvent and identical to the samples mentioned above. The color obtained is compared with the colorimetric scale. A quantity of oil per gramme of rock is thus obtained.

(c) The method can also be used with the solvent contained in the ASTM trap after washing of the sample.

The method can be improved through the use of a photocolorimeter. It should be considered as a semi-quantitative method. The aging of samples used for the colorimetric scale should be noted.

4

natural rock radioactivity measured in the laboratory

41. NATURAL RADIOACTIVITY OF DRILL CUTTINGS

The natural radioactivity (β or γ) of rock samples from outcrops and drill cuttings has been measured either as rocks are collected or on series of preserved samples. Since petroleum is not radioactive, the objective of these measurements is to find certain radioactive markers which show up very clearly during the recording of Gamma Ray radiation in the well. The advantage of measuring natural radioactivity rather than that of drill cuttings lies in the possibility of locating certain layers without stopping the drilling.

This process was developed by the *Institut Français du Pétrole* in 1953 at the request of the Geological Department of the *Société Nationale des Pétroles du Languedoc Méditerranéen* and also in order to study heavy minerals and surface samples in certain sludges.

The apparatus utilized is shown in Fig. 41.1a, b, c and consists of:

(a) A normal Geiger-Müller counter, or a Geiger-Müller counter with a window or a photoscintillator.
(b) A lead shield acting is a screen against cosmic rays in order to reduce the specific movement.
(c) An appropriate counter-integrator.

The sample can be:

(a) Either roughly ground or in the form of 100 to 200 g of cuttings placed around the counter (Fig. 41.1a).

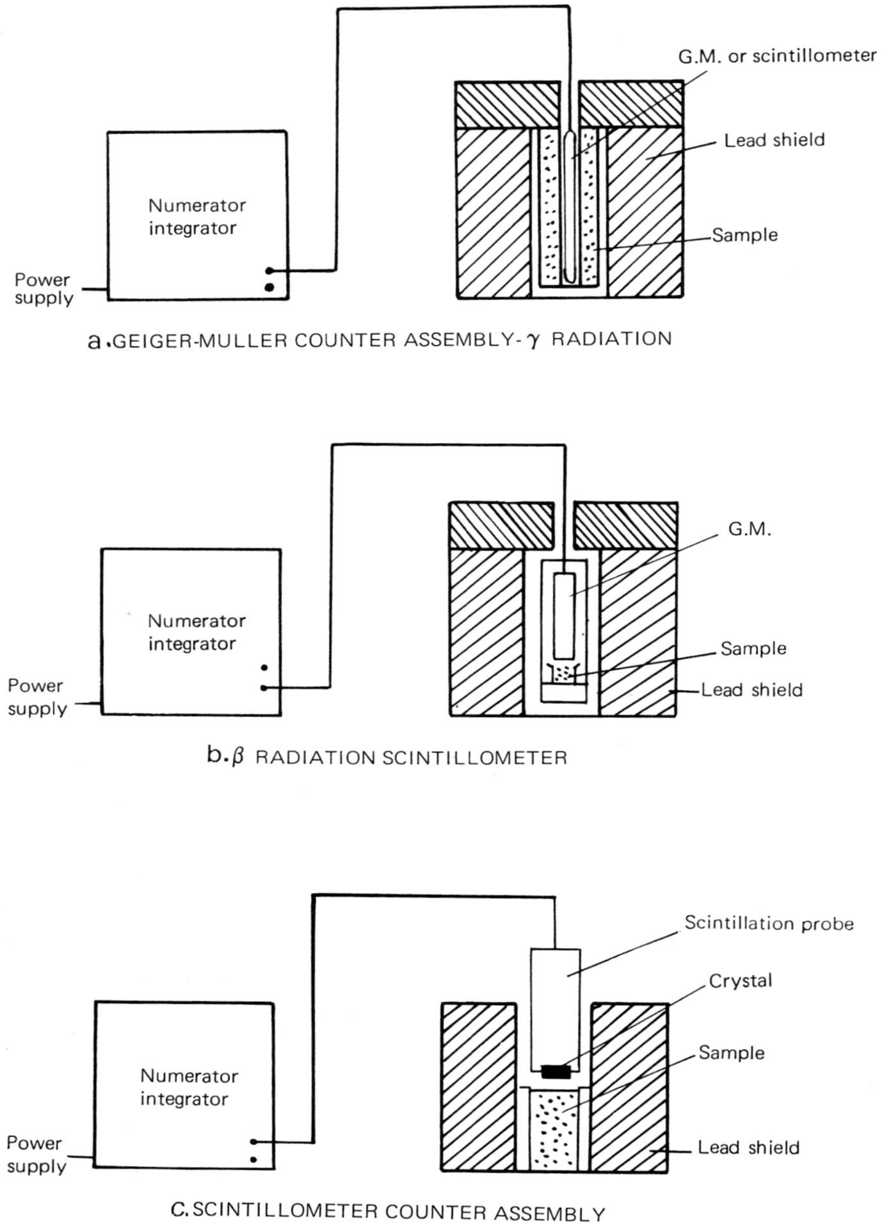

Fig. 41.1. Diagram of assemblies with GM or scintillation counters.

(b) Or finely ground and placed in front of the counter window (Fig. 41.1b, c) or in front of the crystal of a scintillation counter (Fig. 41.1c)

From the experimental point of view:

(a) The number of Gamma Ray pulses in a given period of time is counted.
(b) The result is related to a constant quantity of rock.
(c) A diagram is plotted of the number of impacts for a constant quantity of rock in terms of depth.

Measurement errors are due to:

(a) Counting of pulses distributed according to laws of probability.
(b) Variations in specific movement.
(c) Weighing operations.
(d) Time measurements
(e) Impurities.

The first two are the most important.

42. NATURAL RADIOACTIVITY OF CORES

The natural Gamma Ray radioactivity of rocks is measured in the laboratory on cores recovered from boreholes in order to obtain data without destroying the samples under study. The apparatus shown in Fig. 42.1, which was developed by *Core Laboratories,* consists of:

(a) A conveyor belt upon which are placed the core or the fragments end to end. The core is placed upon a trolley, and an operator at one end of the conveyor belt places the samples on the belt. Another operator at the other end recovers the rock fragments and places the core on another trolley.

(b) A lead tunnel for canceling out or reducing to a great extent the radioactivity due to the specific movement. A window allows natural radiation to reach the Geiger-Müller counter or a photoscintillator.

(c) A photoscintillator or a thin-window Geiger-Müller counter.

(d) Electronic circuits: power supply, amplifier, counter, integrator and millivoltmeter recorder.

The natural radioactivity log is recorded in such a way for sensitivity, apparatus time constant, speed of log and speed of conveyor belt that the recording laboratory log for the core can be compared directly with the Gamma Ray obtained in the well.

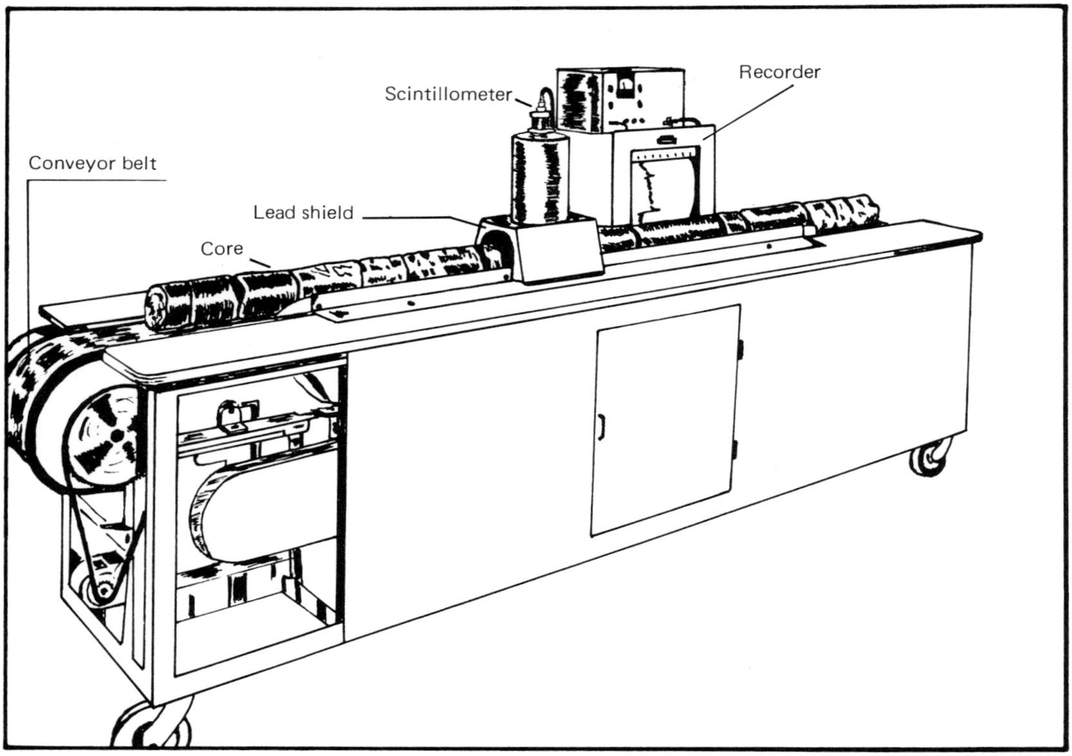

Fig. 42.1. Core-gamma surface log (from *Corelab*).

This laboratory measurement of Gamma Ray radioactivity in rocks is carried out before all other core measurements. It can also be carried out on samples from which the fluids have been removed by distillation.

This measurement does not destroy the sample and can be performed quickly, i.e. 60 min for an 18 m core. The diagram can therefore be compared immediately with results from neighboring wells.

43. EXAMPLES OF THE INTERPRETATION OF NATURAL RADIOACTIVITY MEASURED ON CORES

Amongst the practical applications are the following:

(a) It facilitates correlations and therefore makes it possible to situate productive zones.

(b) It makes it possible to drill cores with precision in thin layers.
(c) It serves as a guide for the core drilling zone.
(d) It makes it possible to eliminate unnecessary coring.
(e) It makes it possible to identify and exactly situate lost core sections.

The diagrams below from *Core Laboratories Inc.* are real cases from the laboratory.

Figure 43.1. Glorietta Formation, Andrews County, Texas: Core analysis was carried out up to 5 942′ (1 811 m). It showed probable production. Comparison of this analysis with that for a neighboring well seemed to show that the productive zone had not actually been reached. Natural radioactivity was recorded in the laboratory on cores and by comparison with Gamma Ray recordings in the neighboring well the following was obtained:

(a) Good correlation between the two results.
(b) The necessity for coring for approximately another fifty feet (15.24m).

The next core actually reached the reservoir.

Figure 43.2. Grayburg Formation, San Andrews, Andrews County, Texas: The correlation of the natural radioactivity diagram and the Gamma Ray diagram showed that there was a difference of 8 feet (2.44 m) between the elevations of the cores and the Gamma Ray log. This piece of information made it possible to explore precisely the thin layers in the formation.

Figure 43.3. McKee Formation, Pecos Country, Texas: Sand was abnormally present in a first core collected in the interval 5 380-5 430 feet (1 639.8-1 655 m). Through correlation it was possible to situate the core exactly.

Figure 43.4. San Andres Formation, Ector County, Texas: Natural radioactivity was measured in order to see if the roof of the formation has actually been touched (it was very visible on the Gamma Ray). In this case continued coring was not necessary.

Figure 43.5. Strawn Reef Formation, Nolan County, Texas: Natural radiation was measured on cores after extraction of the fluids they contained (whole core analysis) by distillation. There are excellent correlations with Gamma Ray logs.

Figure 43.6. Scurry County, Texas: This is a typical example of the way in which the correlations between Gamma Ray and natural core radioactivity diagrams can be used for precisely situating lost core sections. In this particular case the results were used for completion.

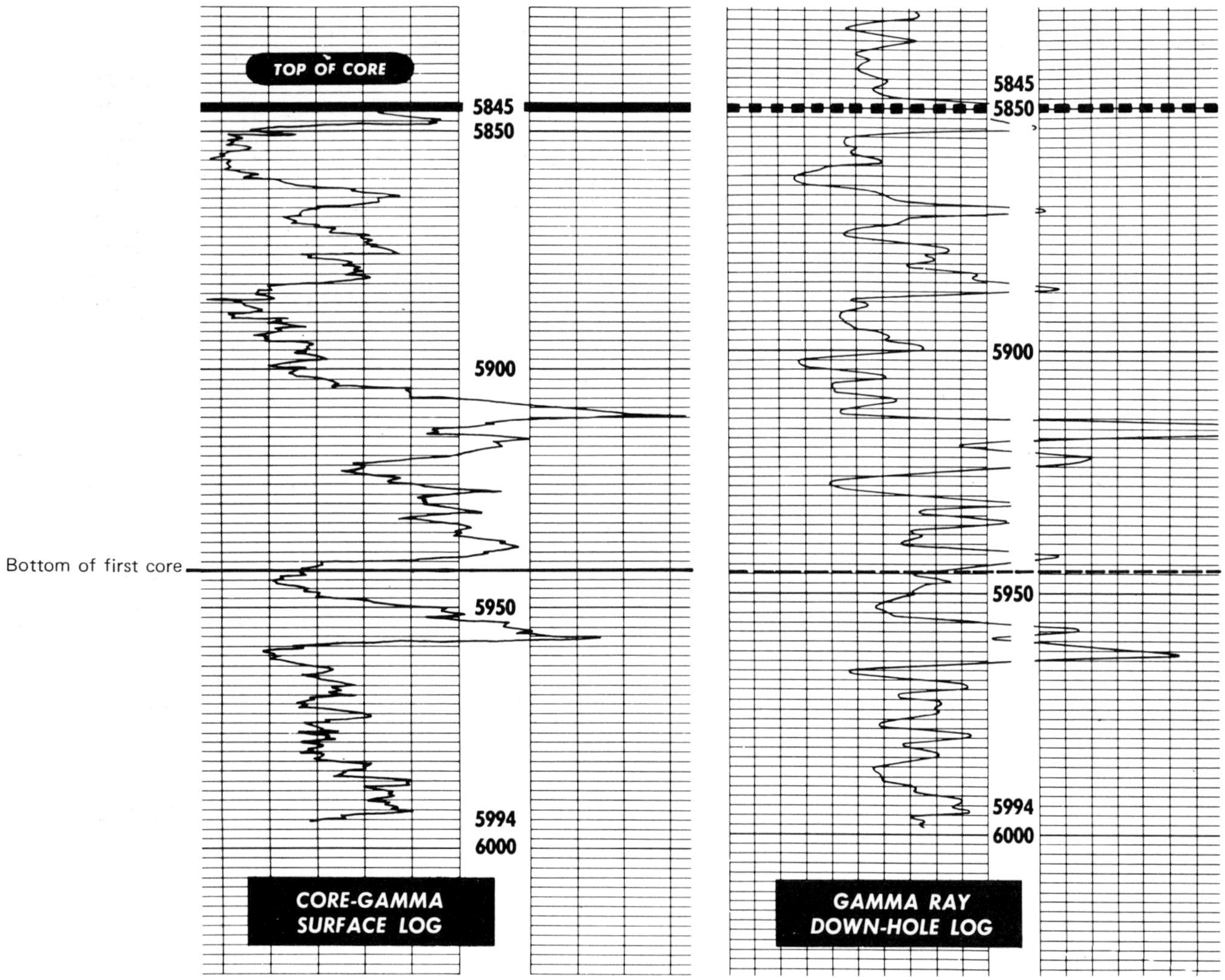

Fig. 43.1. (from *Corelab*).

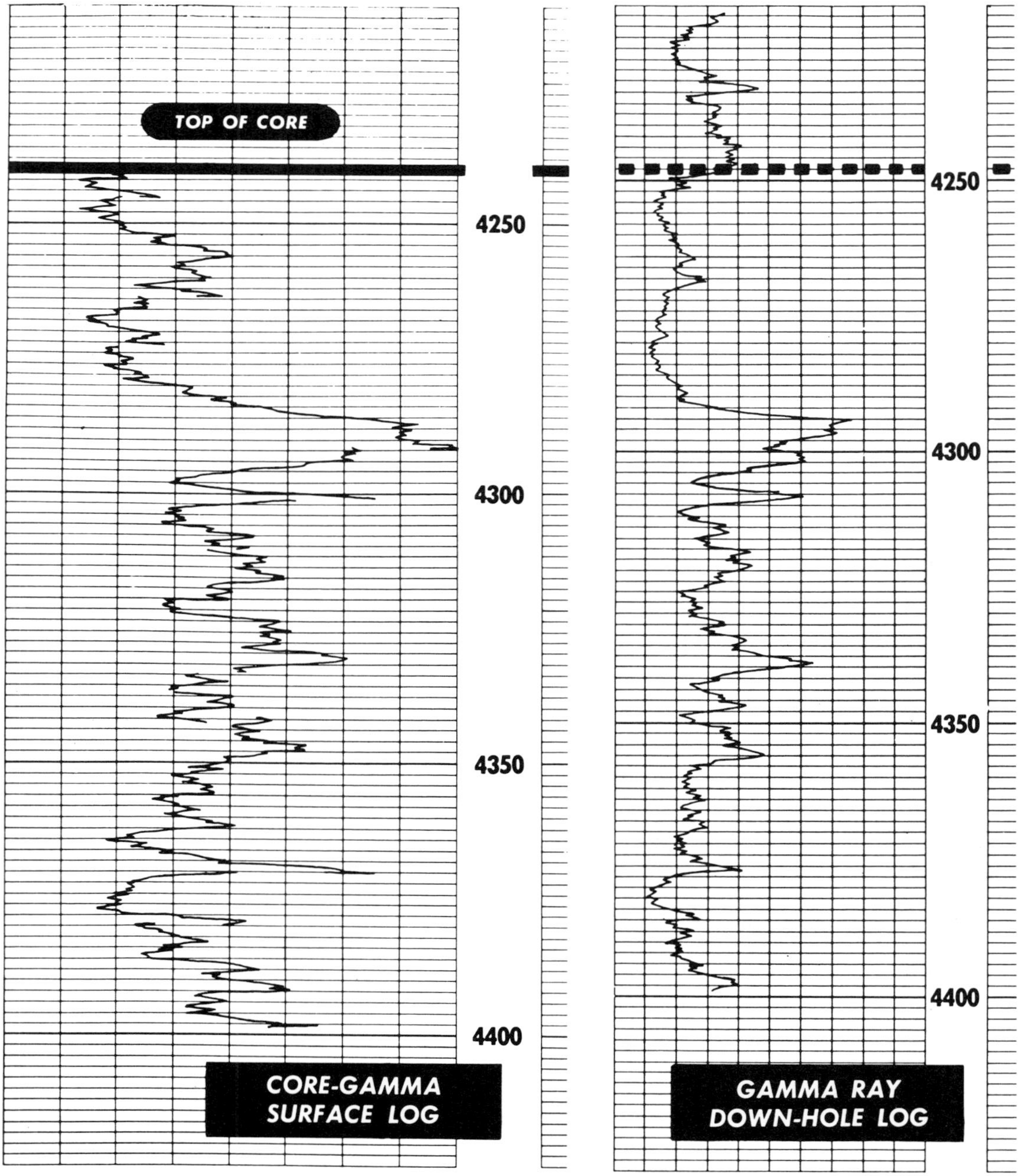

Fig. 43.2. (from *Corelab*).

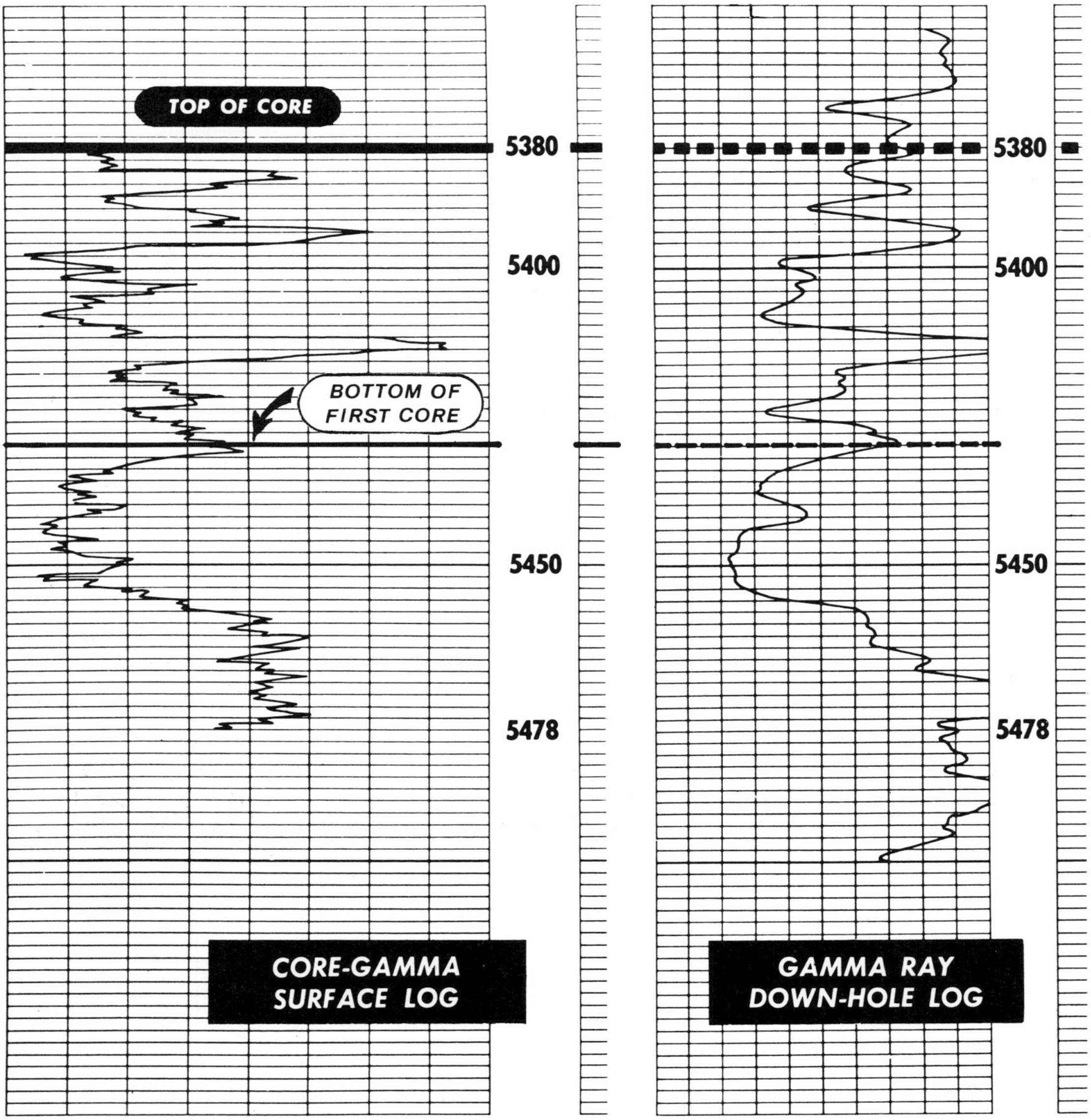

Fig. 43.3. (from *Corelab*).

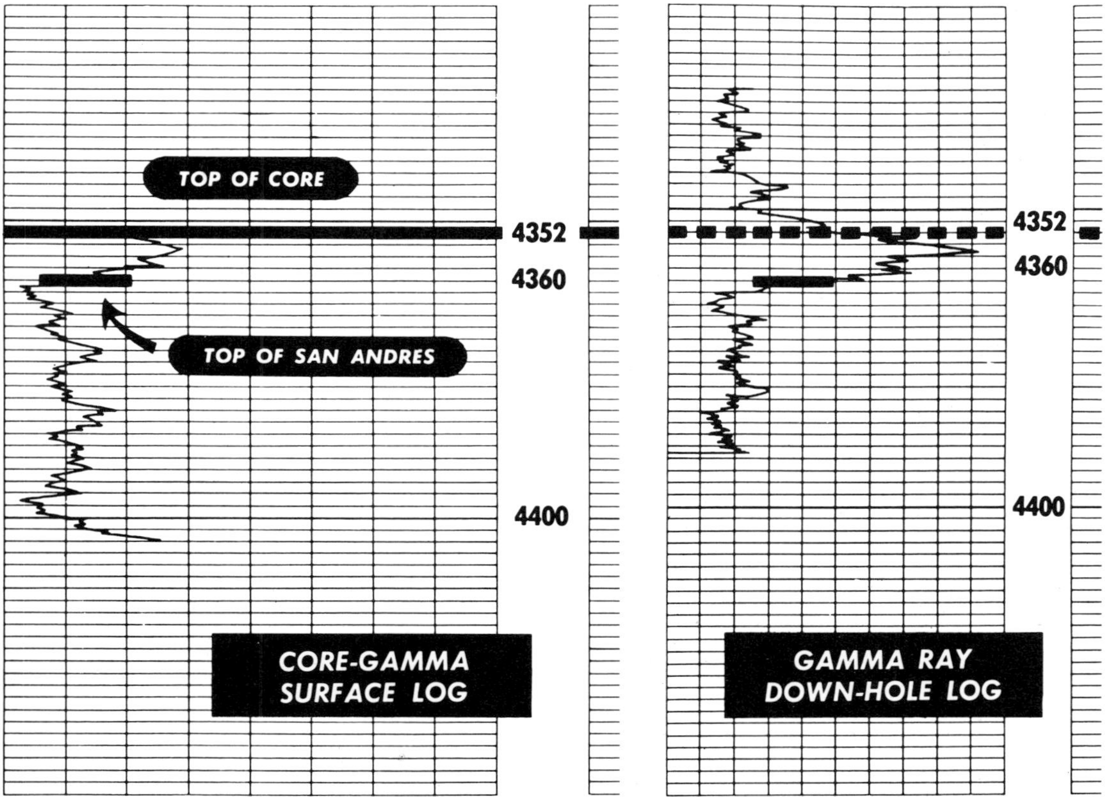

Fig. 43.4. (from *Corelab*).

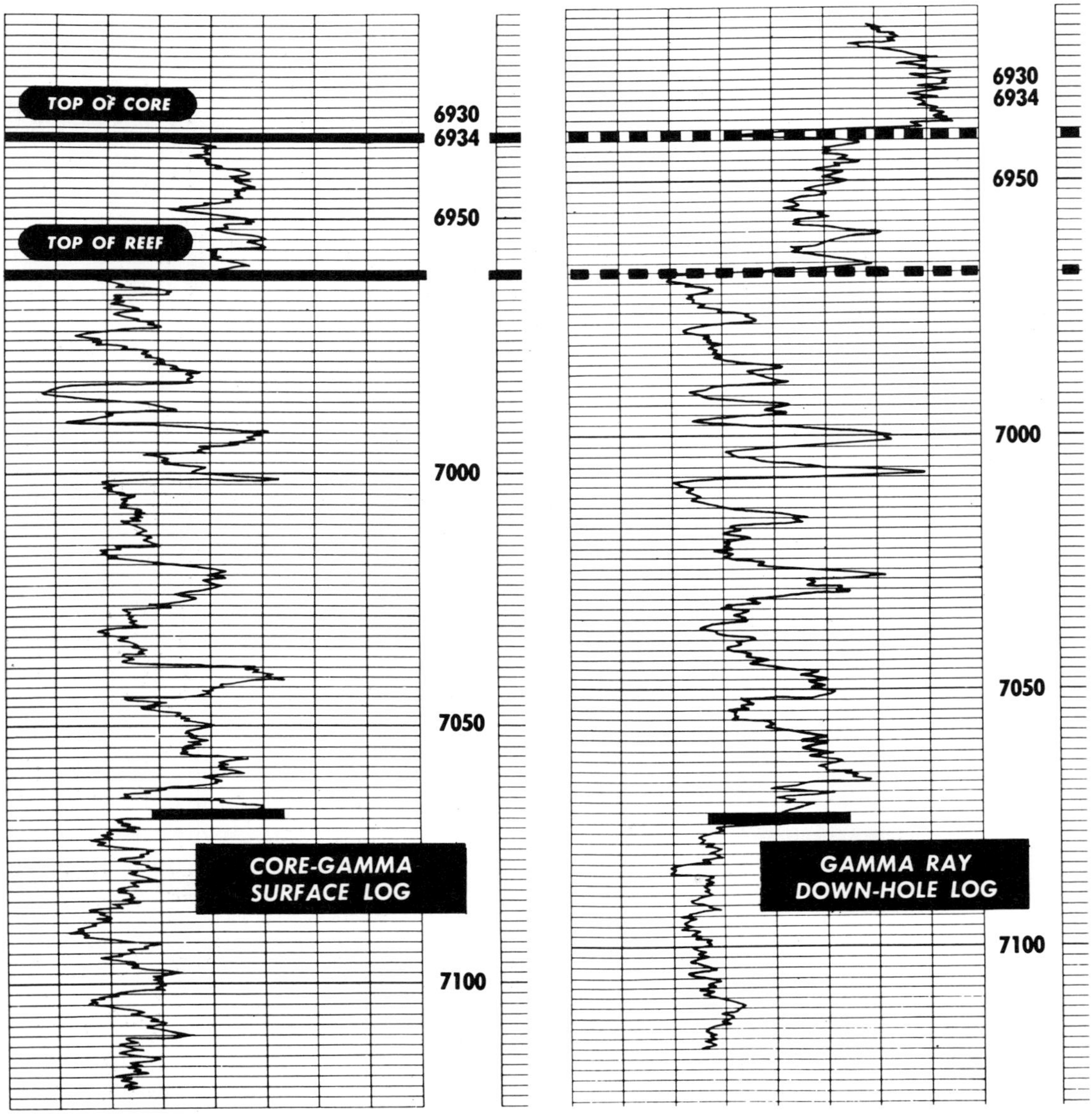

Fig. 43.5. (from *Corelab*).

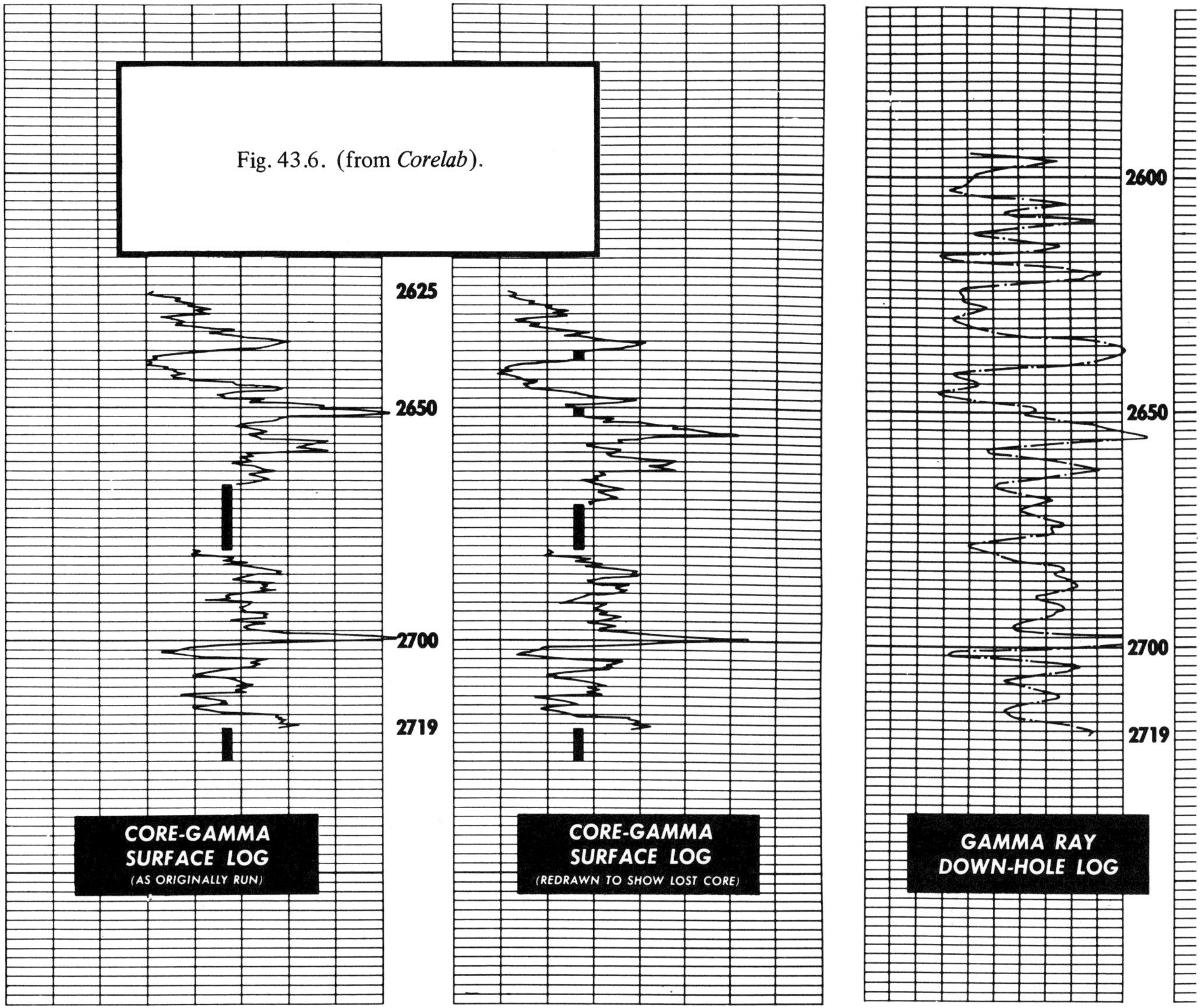

Fig. 43.6. (from *Corelab*).

5

core analysis and interpretation of results

51. GENERAL REMARKS. CLASSIFICATION OF CORE ANALYSIS

Core analysis can be carried out in the laboratory in various ways:

(a) Depending upon the various measurements to be made.

(b) Depending upon the sizes of the samples received or studied.

(c) Depending upon the use to be made of the results for reservoir study needs (special core analysis studies).

The following classification can therefore be made:

51.1. Core analysis

A. Conventional core analysis with small plugs:

(a) Partial analysis with fresh or extracted samples:
. Measurement of air permeability.
. Measurement of porosity.

(b) Complete analysis with fresh or preserved samples:
. Measurement of air permeability.
. Measurement of porosity.
. Measurement of gas, water and oil saturation.

B. Whole core analysis with large or fresh cores:

(a) Measurement of horizontal permeability.
(b) Measurement of porosity.
(c) Measurement of gas, water and oil saturation.

51.2. Special core analysis

(a) **Capillary pressure:**
. By the restored state method.
. By mercury injection, evaporation or centrifuging.
(b) **Liquid permeability.**
(c) **Relative gas/oil permeability.**
(d) **Relative water/oil permeability.**
(e) **Water flood susceptibility test.**
(f) **Formation factor and resistivity ratio.**
(g) **Various measurements:** wettability, etc.

Paragraphs b, c, d, e, f and g are not covered by this course. These measurements are, however, carried out on the basis of core analysis results for a limited selection of samples. This selection is made in terms of permeability or porosity or of the function $\sqrt{k/\phi}$. These studies are necessary for the study of reservoir problems and make it possible to predict field behaviour in time and to estimate reserves. They are made subsequently in specialised laboratories.

52. CONVENTIONAL CORE ANALYSIS

52.1. Removal of cores

A core should be considered as a piece of costly evidence which should be investigated according to a carefully thought out general plan. After being removed from the core barrel, a file card is filled out with the following data:

(a) Footage intervals of the core.
(b) Length collected and inventory of fragments (the lost pieces are arbitrarily placed at the foot of the core).
(c) Detailed geological description and drawing.
(d) Description of shows in impregnated zones (gas, oil, fluorescence).

In the case of saturation measurements, **the core should never be washed** but be wiped off with **clean** rags.

The fragments reserved for immediate analysis are collected and sent to the laboratory of the work site or preserved for later shipment and analysis in a central laboratory which is often far from the work site.

Fragments intended for analyses for determining only physical properties of the reservoir are placed in boxes, well identified and sent to the geological laboratory.

A core analysis laboratory at the work site, or the central laboratory if the distance is not too great, should supply rapid results for the guidance of the rest of the drilling program, i.e. zones with probable oil or water production, drill stem tests to be carried out immediately, transition zones, continued drilling, completion, etc.

52.2. Sampling. Collection of samples and preservation

Sampling varies with the prospection companies. In general the best frequency of collection for conventional analysis with small plugs is:

(a) Three samples per meter for fresh samples for permeability, porosity and saturation measurements. However this frequency is sometimes raised to four samples per meter.

(b) Four samples (H and V) per meter for permeability and porosity measurements only.

For complete analysis, a representative sample (either per third or quarter of a meter) is collected and should be as homogeneous as possible. If there are changes in the sample, each type of sample is collected.

Formations which are not visibly productive or which consist of marls or clays are not analysed and there is only sampling at long intervals.

In the case of whole core analysis the entire core is analysed.

Sampling for side wall core analysis is carried out according to data from other logs. Each sample collected is analysed as far as this is possible.

The cores are carefully protected when core analyses, and in particular saturation measurements, are not made immediately after removal from the core barrel. Of the conventional methods used the following should be mentioned:

(a) Quick freezing method: the cores are stored in a dry ice chamber.
(b) Wrapping of cores with thin sheets of aluminium and paraffin coating.
(c) Use of plastic sacks which may be joined.
(d) Use of fitted boxes.
(e) Use of sealed tubes.

The first two methods give excellent results and the next two are very acceptable in the usual cases. The fifth method is a very special one.

52.3. Conventional core analysis

Two cases can be considered:

Partial analysis: Porosity and air permeability measurements only for exposed and not fresh or extracted samples.

Complete analysis: Porosity measurements, air permeability measurements, saturation measurements: water, gas and oil. This complete analysis can only be carried out for fresh or well preserved samples. In this case the method which seems to give the best results is the so-called fluid summation method.

52.31. Partial core analysis

Air permeability and porosity measurements are made only on samples whose shape is as perfectly geometrical as possible as we have seen for air permeability measurements. The samples are cut either as cylinders ($\phi = 2.3$ cm, $L = 2.5$ to 3.0 cm) or in cubes of 2 cm.

After the samples have been carefully washed in solvent, the following measurements are made:

(a) Air permeability either with a variable head permeameter or with a constant head permeameter.

A cubic sample gives horizontal and vertical permeabilities or, if not, two cylindrical samples must be collected.

The air permeability values are corrected for the Klinkenberg effect by means of laboratory tables so as to obtain k_1 and liquid permeability.

(b) Porosity either by immersion (see course on porosity) which immediately gives the true mass densities or by gas expansion or by some other method, but the first probably gives the best results.

52.32. Complete core analysis by summation of fluids (Method, Core Laboratories Inc., US Patents 2 282 654, 2 345 535 and 2 361 844)

The analysis is carried out on a fresh core sample, wiped with a clean rag (never washed with water), 8 to 10 cm thick and with good representativity. The fresh sample is divided into three parts (Figs. 52.321 and 52.322).

(a) A cube (for horizontal and vertical permeability) or a cylinder (horizontal permeability) is collected from one part.

(b) One part of approximately 30 to 40 g is used for determination of gas saturation.

(c) One part selected as far as possible from the core center is coarsely ground: approximately 125 g. It is used for determination of oil and water saturation through the atmospheric pressure distillation method.

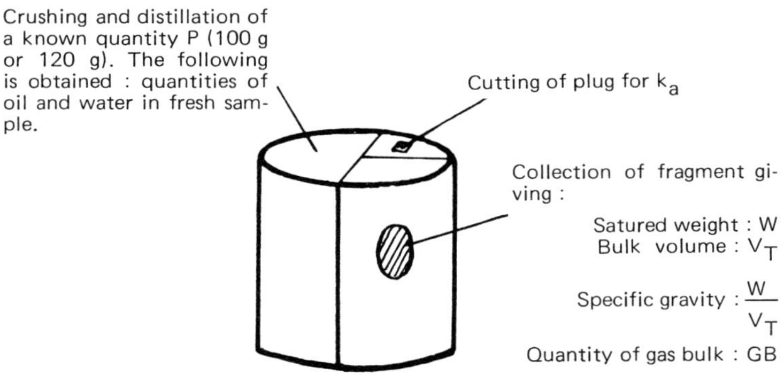

Fig. 52.321. Division of fresh sample for summation of fluids.

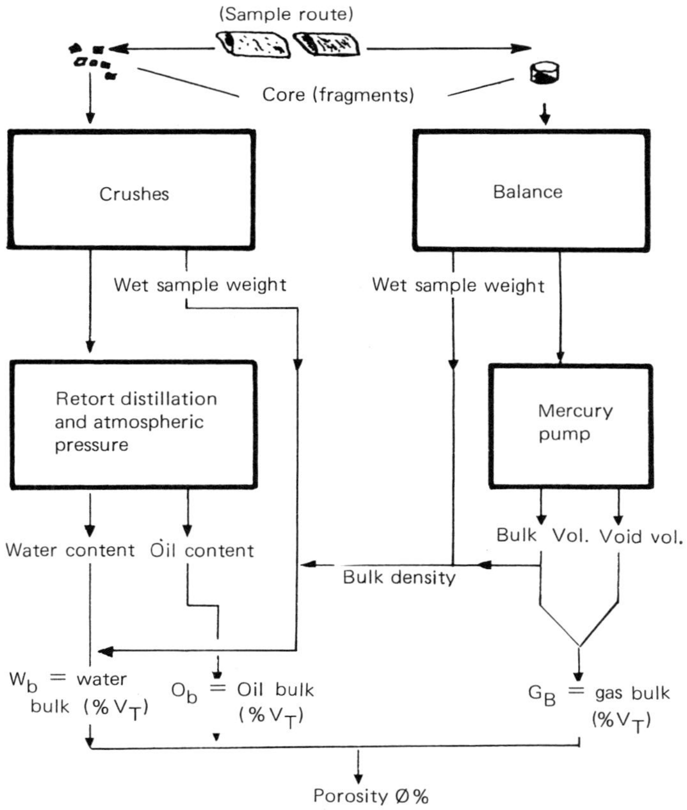

Fig. 52.322. Conventional core analysis by summation of fluids procedures.

Air permeability: A cube (k_H and k_V) or a cylinder (k_H) is collected in water, washed with a combination of a Soxhlet apparatus and a centrifuge and then dried in an appropriate manner. Air permeability is measured by means of a variable or constant head permeameter. This sample is used for measuring permeability and it is preserved.

Measurement of porosity and saturation: The second part, which is lightly rounded off with a hammer (sharp angles should be avoided), is first weighed (approximately 30 g). Let W be its fresh weight.

Its total volume V_T is determined with a mercury pump. The (bulk) density is as follows:

$$\delta = \frac{W}{V_T} \qquad \text{(Eq. 52.321)}$$

The volume of gas contained in this sample is obtained by mercury injection in the following manner: at 34.5 bars the mercury invasion I_1 is read; at 69 bars the new mercury invasion I_2 is read (I_1 and I_2 after correction of pump). It can be shown statistically that the total volume of gas in this fresh sample is $I_2 + (I_2 - I_1)$. Hence the quantity of gas G_b contained in the sample (in percentage) is:

$$G_b = \frac{I_2 + (I_2 - I_1)}{V_T} \cdot 100 \qquad \text{(Eq. 52.322)}$$

If the sample is clayey or partially dried (and G_b is therefore very high) it is preferable to make a single injection at 51.7 bars i.e. I'_1 (after correction). We have:

$$G_b = \frac{I'_1}{V_T} \cdot 100 \qquad \text{(Eq. 52.323)}$$

The third part intended for the determination of oil and water quantities is coarsely ground (bean size). The center of the core is used. A constant weight (of the order of 125 g) is used for retort distillation. The weight P and the bulk density of the sample is known and the volume of the distilled sample is calculated: $V = \frac{P}{\delta}$.

At the end of nine minutes (at the end of which time the clays are still not dehydrated in the oven used) it is possible to read directly the volume of water collected, "Init._W", which was contained in the pores.

At the end of twenty-five minutes (when the oven temperature is of the order of 650° C) it is possible to read the total collected oil volume, "OBS_o".

This observed oil volume should be corrected after cracking. This correction is made after sampling: observed (or measured) oil in terms of real oil volume. From this the true value for the quantity of oil "CORR_o" contained in the sample follows.

Consequently, volume V of ground sample weighing 125 g contains:

$$\text{Water} = W_b = \frac{\text{“Init.}_W\text{”}}{V} \cdot 100 \qquad \text{(Eq. 52.324)}$$

$$\text{Oil} = O_b = \frac{\text{“CORR}_o\text{”}}{V} \cdot 100 \qquad \text{(Eq. 52.325)}$$

Porosity, ϕ, is therefore the sum of fluids contained in the sample expressed as a percentage of the volume:

$$\phi\% = G_b\% + W_b\% + O_b\% \qquad \text{(Eq. 52.326)}$$

For calculating saturations we have:

$$\text{Oil saturation} = S_o = \frac{O_b}{\phi} \cdot 100 \qquad \text{(Eq. 52.327)}$$

$$\text{Water saturation} = S_{TW} = \frac{W_b}{\phi} \cdot 100 \qquad \text{(Eq. 52.328)}$$

It should be noted here that total water saturation is connate water plus water of invasion. A factor (*SF*) was established empirically depending upon the values of ϕ and G_b which makes it possible to determine connate water saturation S_{cw}:

$$S_{cw} = S_{TW} \cdot (SF) \qquad \text{(Eq. 52.329)}$$

The results are presented in the form of tables or curves (Fig. 53.531, 53.532, 53.533 and 53.534 from *Core Laboratories Inc.*). We have:

(a) Sample number — 1 to n.
(b) Sample elevation — in meters or feet.
(c) Air permeability — k_a in mD (horizontal).
(d) Porosity — ϕ as percentage.
(e) Residual oil saturation — S_o in percentage of pore volume.
(f) Total water saturation — S_{TW} is percentage of pore volume.
(g) Observations — fissures (F), vugs (V).
(h) Probable production (water, oil, gas or nothing) in some cases.
(i) Porosity, permeability, oil saturation, water saturation curves.
(j) Geological profile.

In preceding chapters we have seen:

(a) The advantages, disadvantages and accuracy of the various apparatus and methods used.

(b) In addition, the fluid summation method, as well as giving excellent results, also makes it possible to **completely** analyse an 18 m core, i.e. 54 levels in a period of the order of 6 to 8 hours.

52.4. Side wall core analysis

Side wall core analysis makes it possible to collect small samples from the side walls on the basis of other logs (electric, mudlogging or well geology). This core drilling makes certain savings possible without, however, supplying as much data as actual core drilling. Thus data can be obtained for the reservoir, while porosity, permeability and saturation measurements can be made on very small samples ($\phi \simeq$ 23-25 mm, $L \simeq$ 15-40 mm) which are generally relatively consolidated.

The method used is again the fluid summation method. Two variants are possible:

52.41. Distillation method (retort method) (at atmospheric pressure)

The fresh sample is cleaned with a knife and divided into two parts:

(a) The first part is used for air permeability measurements after coating with wax (Fig. 29.52).

(b) The second part is treated as above but with an appropriate material:

Porosity is determined by fluid summations

Gas saturation is determined by mercury injection.

Oil and water saturation are determined by distillation in a small retort (Fig. 52.411).

Mercury, oil and water are collected.

This very quick method gives good results with samples which are representative to varying degrees of the formation and are weathered.

52.42. Extraction with pentane (pentane cut method)

The fresh sample is again divided into two parts:

(a) The first part is treated as above for permeability measurement.

(b) On the second part of the side wall sample:

The weight is determined by weighing.

The quantity of gas is determined by mercury injection.

The volume is determined with a mercury pump.

The sample is ground and placed in 100 cm^3 of pentane.

The pentane is evaporated (heating in a bath of T° = Const. (Fig. 52.421).

The residual oil volume is noted and the latter figure is subjected to an experimental correction (Fig. 52.422).

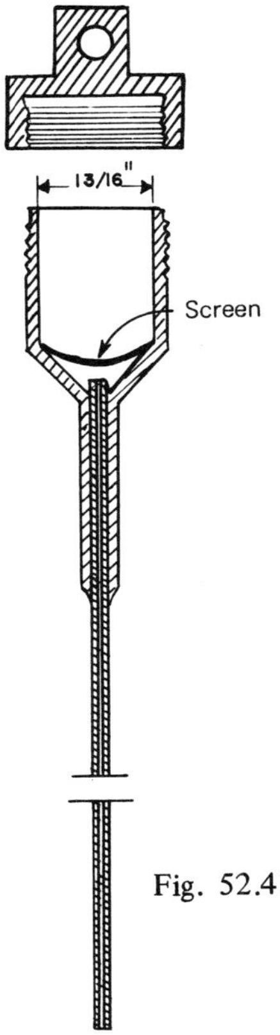

Fig. 52.411. Retort cup. Side wall sample.

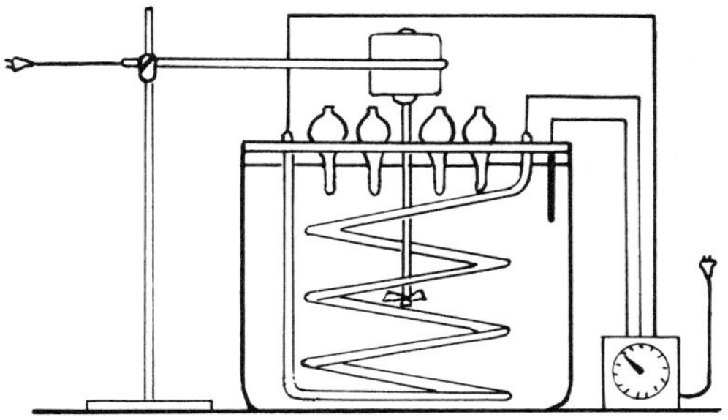

Fig. 52.421. Pentane evaporation apparatus.

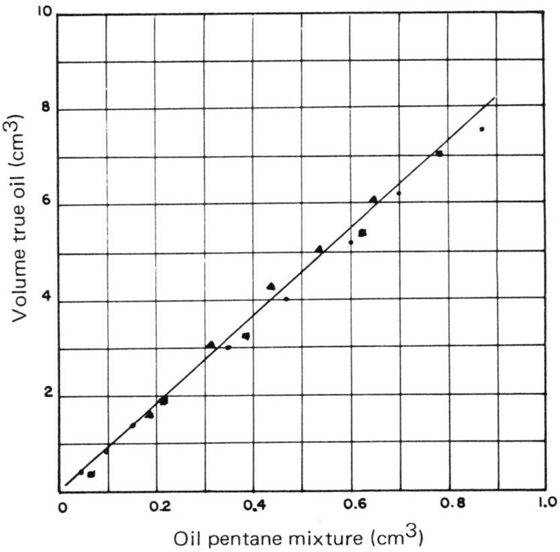

Fig. 52.422. Pentane evaporation data. Water bath temperature: 160° F.

We then have:

(a) The gas measured above.
(b) Total oil.
(c) The water is calculated with a material balance.
(d) Porosity is determined by fluid summation.

This method is no doubt more complicated than the distillation method but it is often quicker.

The results are presented as in the preceding method and the measurements are completed by a lithological description and observations of fluorescence.

52.5. Whole core analysis

This analysis is made on the whole core. Each core fragment is treated as a single sample. Small fragments are kept for the collection.

The core reconstructed from its fragments may be placed on a conveyor belt for scanning by a Geiger counter or a photoscintillometer for measurement of natural radioactivity. This natural radioactivity diagram made in the laboratory when compared with that obtained in the well (Gamma Ray) makes it possible to situate the parts of the core or certain markers.

Color photos of the core may be made.

The following method of analysis (Fig. 52.51) is recommended in cases of cracked or vuggy limestone:

Determination of the quantity of gas G_b contained in each fragment:

(a) The fragment is weighed with its fluids: P_1.

(b) The fragment is placed in a vacuum and then saturated with deaerated water and brought under pressure for 45 minutes.

(c) The fragment is removed and weighed again: P_2.

(d) The increase is weight $P_2 - P_1$ corresponds approximately to the volume of gas present.

Determination of quantities of water and oil contained in each fragment:

After the weighing of the water saturated fragment (P_2), vacuum distillation is carried out. The sample is heated for two hours at 120° C and two hours at 230° C. If there is visibly little porosity it is heated for four hours at 230° C. During distillation pressure is of the order of 20 to 30 mm of mercury (absolute).

The fluids are collected in graduated test tubes and placed in a cooling solution (methyl alcohol + carbon dioxide ice – $T° \simeq -60°$ C):

(a) Collection of water: water in pores plus water corresponding to gas. Since the samples are limestone or dolomite there is no reason to fear the clay dehydration phenomenon.

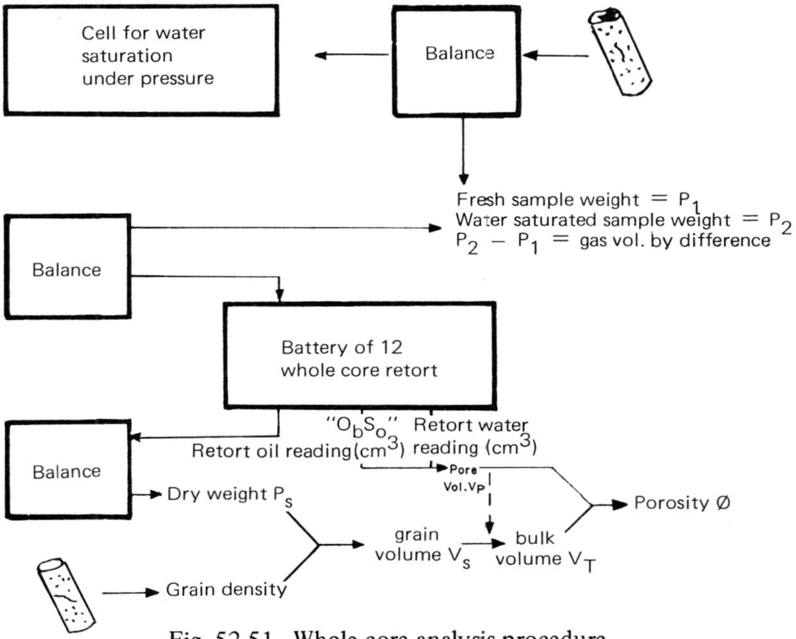

Fig. 52.51. Whole core analysis procedure.

(b) Collection of oil: a corrective factor should be applied because of cracking.

The sample is weighed after removal from the oven $= P_{dry}$

Loss of weight $= P_2 - P_{dry} \approx$ pore volume.

The total volume of the fragment is calculated as follows:

$$\text{Total volume } V_T = \frac{\text{Dry weight}}{\text{True density}} + (\text{Weight loss}).$$

The average grain density values are:

2.82 for dolomites,
2.72 for limestones.

Calculation of percentages of gas, oil and water contained in the fragment:

$$\%\text{ gas} = G_b = \frac{P_2 - P_1}{V_T} \cdot 100 \qquad \text{(Eq. 52.51)}$$

$$\%\text{ oil} = O_b = \frac{\text{corrected quantity of oil collected}}{V_T} \cdot 100 \qquad \text{(Eq. 52.52)}$$

$$\%\text{ water} = W_b = \frac{\text{quantity of water in sample}}{V_T} \cdot 100 \qquad \text{(Eq. 52.53)}$$

Hence

$$\text{Porosity: } \phi = G_b + O_b + W_b \qquad \text{(Eq. 52.54)}$$

Residual oil saturation:

$$S_o = \frac{O_b}{\phi} \cdot 100 \qquad \text{(Eq. 52.55)}$$

Residual water saturation:

$$S_w = \frac{W_b}{\phi} \cdot 100 \qquad \text{(Eq. 52.56)}$$

Measurement of air permeability

The measurement is made on a core fragment after vacuum distillation. Horizontal permeability is determined in two perpendicular directions. The ends of the fragment are soaked in liquid so as to eliminate air flow from the fragment ends.

The fragment is placed between two screens and between two rubber half shells (Fig. 29.33) which are held and squeezed by a hydraulic system (Fig. 29.31).

Permeability is measured with a constant head permeameter. A geometric correction factor takes into account total length L of section A and the length not covered by wax G (Fig. 29.32).

This method is recommended for limestone or dolomite deposits which are very cracked or vuggy and where conventional analysis is not representative. The results are presented as for conventional analysis.

53. INTERPRETATION OF CORE ANALYSIS

The following discussion applies especially to the fluid summation method used for conventional core analysis and whole core analysis.

53.1. General remarks

(a) Each sample analysed comes from a wiped (and not washed) core which has been preserved for saturation measurement.

(b) Three (or four) samples per meter are studied in the reservoirs.

(c) Analysis is carried out on representative samples.

(d) The sample is analysed by the fluid summation method (completely described above).

(e) Air permeability is measured on the horizontal plane. A cube is collected for determination of k_h and k_v.

(f) The total water saturation measurement is corrected for filtrate invasion. Tables (established empirically) give connate water saturation.

(g) Analysis, study and results are confidential.

53.2. Basic considerations

(a) There is a fundamental condition, i.e. the superposition from the bottom up of water, oil and gas.

(b) It is necessary to establish for the field the minimal oil and maximum water saturations allowing oil and gas production (anhydrous production).

(c) If saturation at one point is not as predicted, the corresponding zone should be limited to this point and the gas/oil/water contacts should be determined.

(d) Good levels should be carefully considered.

53.3. Qualitative interpretation: relation between depth and permeability, porosity and lithology

Core analysis makes it possible to:

(a) Mark the permeable zones and vertical permeability barriers.

(b) Carry out statistical studies on permeability values as well as porosity, either with individual values or average values per interval.

(c) Establish permeability profiles and show distribution in relation to depth.

(d) Determine whether oil is present or not.

(e) Establish statistical and structural correlations as well as correlations with other logs.

53.4. Quantitative interpretation: distribution of fluids and different types of production

Core analysis makes it possible to:

(a) Define storage and production capacities (ϕh; Kh).

(b) Estimate connate water saturation (oil base mud).

(c) Forecast the type of fluid produced.
(d) Define transition zones.
(e) Determine the position of the contacts between various fluids.
(f) Determine the possibility of first occurrence of water or gas.
(g) Forecast the possibility of water coning or gas coning and vertical permeability.
(h) Forecast probable gas and condensate production.
(i) Determine oil field water and its resistivity.
(j) Determine the true mass density of the rock for the interpretation of logs.

It should be noted that, in certain cases, analysis interpretation is very difficult or even impossible and could even give very different results from those obtained by other methods, such as tests. Amongst the possible errors or anomalies are the following:

(a) Large lithological variations within the bed present difficulties when situating the gas/oil, water/oil contacts.
(b) Invasions of formations either by mud or filtrate lead to test flows which are either very small or null and their interpration gives permeabilities which are quite different from those given by core analysis (water blocking).
(c) Unknown cracks can lead to unexpected production from the point of view of size as well as character.
(d) Imperfect collection reduces the value of core analysis. It is often the best part of the reservoir which is not sampled, and in this case it is necessary to use other data for example from outflow rate logs or electric logs.
(e) Abnormal variations in relative permeability.

It should not be forgotten that the calculations which can be made are not precise and that human error is also possible.

53.5. Examples of interpretation

There are too many typical examples to be given here. Therefore only a few cases of interpretation with diagrams from a course taught at *Core Laboratories Inc.* will be given below.

53.51 FRIO formation: clean sands

There is oil and gas production if $S_{TW} < 50\%$.
There is oil production if $S_o > 8\%$.
There is gas production if $S_o < 2\%$.

Example

k_a(mD)	ϕ %	S_o %	S_{TW} %	Probable Production	Observations
15	22	10	40	Oil	Difference
15	22	2	40	Condensate gas	in
15	22	2	60	Water	fluorescence

53.52. COCKFIELD formation: relatively clean sands

There is oil production if $S_{TW} < 60\%$.
There is oil production if $S_o > 8\%$.
There is gas production if $S_o < 3\%$.

Lithology	k_a (mD)	ϕ %	S_o %	S_{TW} %	Probable production	Observations
Clean sandstone	380	29	10	47	Oil	
Clean sandstone	350	30	2	42	Gas	
Clean sandstone	300	28	2	65	Water	
Clayey sandstone	15	24	3	70	Oil	Clayey
Clayey sandstone	50	24	1	75	Gas	
Clean sandstone	300	28	20	20	Oil	$S_{TW} \approx S_{cw}$
Clean sandstone	400	30	15	40	Oil	
Clean sandstone	500	31	7	50	Water	
Clean sandstone	300	30	2	60	Water	
Clays				Not analysed		

Remarks

In interpretation for a single point, conditions before and after the point under consideration should be taken into account.

Water saturation amplitudes depend upon:

(a) Grain porosity and distribution.
(b) Sample height above water level.

They are controlled by capillary phenomena.

Residual oil saturations of the order of:

5 to 10% correspond to light oil,
14 to 24% correspond to average density oil,
30 to 35% correspond to very viscous oil.

53.53. Examples

First example

The example in Fig. 53.531 from *Core Laboratories* is a particularly interesting one.

(a) The top of the reservoir is located at approximately 4 805 feet (1 464.6 m). From 4 805 to 4 808.50 feet (1 465.6 m) permeability is very low because the formation is clayey (or marly). Oil saturations are relatively high.

(b) From 4 808.5 to 4 829.00 feet (1 471.9 m) we observe:

. An increase in permeability clearly indicating influx into the reservoir towards 4 808.50 feet (1 465.6 m).
. Porosity of between 15 and 20%.
. Fluid saturation indicating that this zone is gas producing. Oil saturation is null, water saturation very low and the production capacity ($k \times h$) makes it possible to forecast good performances.

(c) From 4 830.00 to 4 846.00 feet (1 477.1 m) we have:

. Residual oil saturation and total water saturation making it possible to forecast oil production.
. Permeabilities are high.
. Porosities are excellent.

The reservoir described here is therefore very interesting. In addition the vertical fractures (*VF*) in the gas and oil zones should be noted resulting in a completion problem which will be examined below.

(e) From 4 846.00 to 4 849.00 feet (1 478 m) we have:

. An impermeable marl barrier followed by an impregnated reservoir passage and then another very marly foot. If continuity exists the impermeable barriers will be very useful for completion.
. A clear reduction in oil saturation and the physical characteristics of the reservoir.

(f) From 4 849.00 feet to the bottom we have:

. 2 feet of permeable sandstone reservoir with zero oil saturation but with very high water saturation. This part of the reservoir is situated in the aquifer. The water/oil contact is clearly indicated at the level of 4 849.00 feet.
. The reservoir does not extend below 4 851 feet (1 478.6 m).

(g) From the point of view of well completion :

. The permeability barriers (4 846 feet i.e. 1 477.1 m and below) if they are continuous will probably make production possible without significant water inflow.

CORE LABORATORIES, INC. CORE LAB *Petroleum Reservoir Engineering*

COMPANY	GOOD OIL COMPANY	DATE ON	9-1-52
FILE NO.	UAP-1(FC)		
WELL	SMITH NO. 1	DATE OFF	9-16-52
ENGRS.	EFM-TBG		
FIELD	WILDCAT	FORMATION	DAKOTA "J" SAND
ELEV.	4277' KB		
COUNTY	CHEYENNE, STATE NEBRASKA	DRLG. FLD.	WATER BASE MUD
CORES	DIAMOND		
LOCATION		REMARKS	SERVICE NO. 5

SAND · SHALE · LIMESTONE · DOLOMITE · CONGLOMERATE · CHERT

TABULAR DATA and INTERPRETATION

No Fracture--NF
Vertical Fracture--VF

SAMPLE NUMBER	DEPTH FEET	PERM. MD.	POROSITY %	RESIDUAL SATURATION % PORE SPACE OIL	RESIDUAL SATURATION % PORE SPACE TOTAL WATER		PROD
29	4805.5	0.0	7.5	0.0	68.0	NF	
30	06.5	0.0	12.3	0.0	78.0	NF	
31	07.5	2.5	17.0	0.0	43.0	NF	Gas
32	08.8	59	20.7	0.0	29.0	NF	Gas
33	09.5	221	19.1	0.0	31.4	NF	Gas
34	10.5	211	20.4	0.0	38.7	VF	Gas
35	11.5	275	23.3	0.0	34.7	NF	Gas
36	12.5	384	24.0	0.0	26.2	NF	Gas
37	13.5	108	23.3	0.0	30.9	NF	Gas
38	14.5	147	16.1	0.0	29.2	NF	Gas
39	15.5	290	17.2	0.0	34.3	VF	Gas
40	16.5	170	15.3	0.0	24.2	NF	Gas
41	17.5	278	15.9	0.0	26.4	NF	Gas
42	18.5	238	18.6	0.0	39.8	NF	Gas
43	19.5	167	16.2	0.0	39.5	VF	Gas
44	20.5	304	20.0	0.0	38.0	VF	Gas
45	21.5	98	16.9	0.0	34.3	NF	Gas
46	22.5	191	18.1	0.0	34.8	VF	Gas
47	23.5	266	20.3	0.0	31.1	VF	Gas
48	24.5	40	15.3	0.0	22.9	VF	Gas
49	25.5	260	15.1	0.0	13.9	VF	Gas
50	26.5	179	14.0	0.0	21.4	VF	Gas
51	27.5	312	15.6	0.0	28.8	NF	Gas
52	28.5	272	15.5	0.0	34.8	VF	Gas
53	29.5	395	19.4	6.2	25.3	NF	Trans.
54	30.5	405	17.5	13.1	25.7	VF	Oil
55	31.5	275	16.4	17.7	22.5	NF	Oil
56	32.5	852	17.2	19.8	19.2	VF	Oil
57	33.5	610	15.5	21.9	21.3	VF	Oil
58	34.5	406	20.2	16.3	22.3	VF	Oil
59	35.5	535	18.3	19.7	24.6	VF	Oil
60	36.5	663	19.6	19.4	16.3	VF	Oil
61	37.5	597	17.7	17.5	19.8	VF	Oil
62	38.5	434	20.0	14.0	27.5	VF	Oil
63	39.5	339	16.8	20.8	19.7	VF	Oil
64	40.5	216	13.3	18.1	23.3	VF	Oil
65	41.5	332	18.0	15.6	15.6	VF	Oil
66	42.5	295	16.1	19.3	15.5	VF	Oil
67	43.5	882	15.1	19.2	21.2	NF	Oil
68	44.5	600	18.0	20.6	22.2	VF	Oil
69	45.5	407	15.7	15.3	13.4	VF	Oil
70	4847.5	479	17.8	20.8	14.6	VF	Oil
71	48.5	0.0	9.2	14.1	8.7	NF	
72	49.5	139	20.5	0.0	77.1	NF	Water
73	50.5	135	8.4	0.0	57.2	NF	Water
74	51.5	0.0	1.1	0.0	63.6	VF	

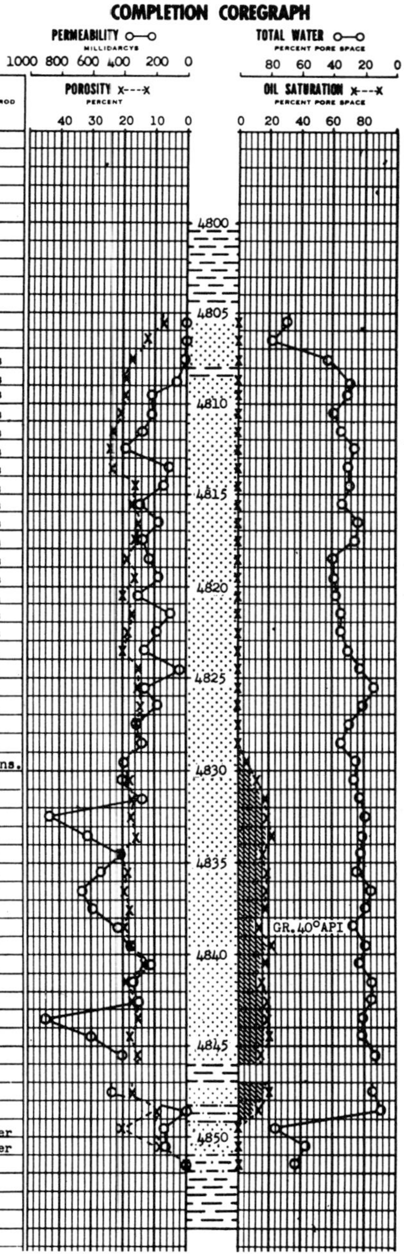

Fig. 53.531. (from *Corelab*).

. There are vertical fractures in the oil and gas reservoir, and if all the energy in the gas cap is to be used while avoiding gas coning the casing should be perforated sufficiently far from the gas/oil contact.

. The best interval for perforation for well production is 4 840-4 843 feet (1 475-2-1 476.1 m).

Second example (Fig. 53.532)

The second example (from *Core Laboratories Inc.*) is also an excellent one. It shows gas/oil and water/oil transiton zones. It concerns a very good reservoir where two feet (61 cm) of perforation will be sufficient to start production. The perforated interval is from 6 448 to 6 450 feet (1 965.35 to 1 965.96 m) and the interval is located 5 feet (1.52 m) from the gas/oil contact and 7 feet (2.13 m) from the water/oil contact.

Third example (Fig. 53.533)

This example (from *Core Laboratories Inc.*) concerns a field of relatively light oil (API gravity = 40°-41°). Residual oil saturation is quite low. The water/oil contact is very clear at 4 690 feet (1 429.5 m). Permeability near the aquifer is very high and there is a high risk of water coning.

The perforated interval is located quite far from the water level at from 4 671 to 4 676 feet (1 423.7 to 1 425.2 m).

Fourth example (Fig. 53.534 from *Core Laboratoires Inc.*)

Core analysis brings out the following:

(a) A gas producing zone: 5 524 to 5 532 feet (1 683.7 to 1 686.1 m).
(b) Gas/oil contact at 5 531 feet (1685.8 m).
(c) An oil producing zone: 5 557 to 5 561 feet (1 686.1 to 1 693.8 m).
(d) After 5 561 feet the production probably consists of water.
(e) The presence of marls reduces vertical permeability in certain cases.

The interval recommended for completion is 5 537 to 5 546 feet (1 687.7 to 1 690.4 m).

54. CALCULATIONS

It should be noted that there is a 'tendency' to do less core drilling. If better data are not available the use of correlations will continue to be valid.

In addition to the immediate interpretations which can be based on them, the use of core analyses make possible some very useful calculations which are very valuable for reservoir studies on first approximations.

CORE LABORATORIES, INC. CORE LAB *Petroleum Reservoir Engineering*

COMPANY: GOOD OIL COMPANY		DATE ON: 6-11-53	FILE NO.: UAP-9(FC)
WELL: FEE NO. 2		DATE OFF: 6-14-53	ENGRS.: JC
FIELD: JACK		FORMATION: COCKFIELD	ELEV.:
COUNTY: HARDIN	STATE: TEXAS	DRLG. FLD.: WATER BASE MUD	CORES: WIRE LINE
LOCATION:		REMARKS: SERVICE NO. 5	

SAND SHALE LIMESTONE DOLOMITE CONGLOMERATE CHERT

These analyses, opinions or interpretations are based on observations and material supplied by the client to whom, and for whose exclusive and confidential use, this report is made. The interpretations or opinions expressed represent the best judgment of Core Laboratories, Inc. (all errors and omissions excepted), but Core Laboratories, Inc. and its officers and employees, assume no responsibility and make no warranty or representations as to the productivity, proper operation, or profitableness of any oil, gas or other mineral well or sand in connection with which such report is used or relied upon.

TABULAR DATA and INTERPRETATION

SAMPLE NUMBER	DEPTH FEET	PERM. MD.	POROSITY %	RESIDUAL SATURATION % PORE SPACE			PROD
				OIL	TOTAL WATER		
36	6431.5	10	18.7	0.0	80.8		COND
37	32.2	76	25.5	2.0	72.9		COND
38	33.5	49	30.4	1.6	72.5		COND
39	34.5	21	30.0	1.7	75.3		COND
40	35.5	710	32.9	1.5	47.4		COND
41	36.5	3520	37.0	1.3	53.0		COND
42	37.5	2340	32.6	1.5	51.2		COND
43	38.5	5070	36.3	1.4	51.2		COND
44	39.5	4620	38.4	1.3	47.9		COND
45	40.5	1420	35.6	0.6	58.1		COND
46	41.5	1230	35.3	0.6	59.5		COND
47	42.5	3080	35.0	1.4	49.4		COND
48	43.5	13700	38.0	9.0	35.8		OIL
49	44.5	13500	38.2	9.7	40.8		OIL
50	45.5	1640	38.2	7.9	30.2		OIL
51	46.5	13300	38.9	12.1	52.8		OIL
52	47.5	4600	35.5	12.1	57.2		OIL
53	48.5	10300	29.9	11.7	41.8		OIL
54	49.5	7980	24.1	12.8	41.9		OIL
55	50.5	9500	32.7	13.4	49.2		OIL
56	51.5	2130	31.0	12.9	55.9		OIL
57	52.5	645	32.9	10.6	55.0		OIL
58	53.5	384	38.9	9.0	63.8		OIL
59	54.5	670	38.2	10.2	61.8		OIL
60	55.5	2820	34.9	10.3	58.7		OIL
61	56.5	5300	35.3	11.0	60.6		OIL
62	57.5	1160	38.1	0.0	69.0		WATER
63	58.5	4970	38.9	0.0	71.2		WATER
64	6461.5	4570	38.1	0.0	81.8		WATER
65	62.5	4540	38.8	0.0	68.1		WATER
66	63.5	622	34.8	0.0	69.8		WATER

COMPLETION COREGRAPH

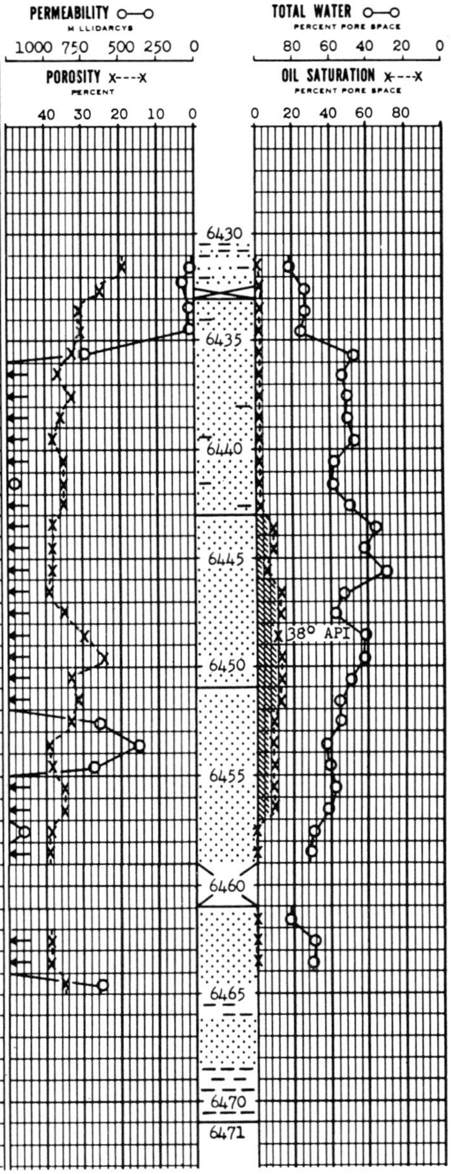

Fig. 53.532. (from *Corelab*).

CORE LABORATORIES, INC. CORE LAB *Petroleum Reservoir Engineering*

COMPANY	GOOD OIL COMPANY	DATE ON	4-6-54
FILE NO.	UAP-8(FC)		
WELL	DAVIS NO. 1	DATE OFF	4-8-54
ENGRS.	HO-CM		
FIELD	HERBERT	FORMATION	WOODBINE
ELEV.	289' DF		
COUNTY	ANDERSON	STATE	TEXAS
DRLG. FLD.	WATER BASE MUD	CORES	DIAMOND
LOCATION		REMARKS	SERVICE NO. 5

SAND LIMESTONE CONGLOMERATE CHERT

SHALE DOLOMITE

These analyses, opinions or interpretations are based on observations and material supplied by the client to whom and for whose exclusive and confidential use, this report is made. The interpretations or opinions expressed represent the best judgment of Core Laboratories, Inc. (all errors and omissions excepted), but Core Laboratories, Inc. and its officers and employees, assume no responsibility and make no warranty or representations as to the productivity, proper operation, or profitableness of any oil, gas or other mineral well or sand in connection with which such report is used or relied upon.

TABULAR DATA and INTERPRETATION

SAMPLE NUMBER	DEPTH FEET	PERM. MD.	POROSITY %	RESIDUAL SATURATION % PORE SPACE OIL	RESIDUAL SATURATION % PORE SPACE TOTAL WATER		PROD
6	4656.5	53	30.1	6.3	52.5		OIL
7	57.5	67	31.9	4.7	51.5		OIL
8	58.5	3.7	24.4	3.7	70.5		OIL
9	59.5	28	26.2	9.2	58.0		OIL
10	60.5	45	28.8	6.6	49.0		OIL
11	61.5	135	30.3	6.3	55.1		OIL
12	62.5	1.5	22.3	8.5	64.2		OIL
13	63.5	81	28.5	10.9	53.0		OIL
14	64.5	923	30.8	10.1	47.1		OIL
15	65.5	107	21.8	6.0	44.5		OIL
16	66.5	272	28.6	6.7	51.1		OIL
17	67.5	212	30.0	11.3	47.3		OIL
18	68.5	11	23.9	10.5	52.3		OIL
19	69.5	84	29.2	10.6	47.3		OIL
20	70.5	150	33.0	9.4	46.6		OIL
21	71.5	1610	33.2	8.7	45.2		OIL
22	72.5	490	28.7	9.4	47.5		OIL
23	73.5	1550	28.2	9.6	44.0		OIL
24	74.5	2080	26.4	9.5	46.2		OIL
25	75.5	665	26.5	12.1	51.8		OIL
26	76.5	1550	29.2	7.9	43.5		OIL
27	77.5	1260	28.6	10.8	48.3		OIL
28	78.5	1550	30.2	10.3	45.1		OIL
29	79.5	580	28.2	11.3	48.0		OIL
30	80.5	1978	29.5	12.2	43.1		OIL
31	81.5	1370	31.7	11.0	47.4		OIL
32	82.5	1128	30.1	9.3	50.5		OIL
33	83.5	1550	27.5	9.8	43.3		OIL
34	84.5	1370	23.7	6.3	39.3		OIL
35	85.5	2600	26.7	6.8	42.0		OIL
36	86.5	1650	30.1	10.3	38.5		OIL
37	87.5	2860	30.3	12.5	40.2		OIL
38	88.5	2490	28.4	14.1	35.2		OIL
39	89.5	2130	27.3	11.7	47.3		OIL
40	90.5	3370	32.5	0.6	75.6		WATER
41	91.5	4720	29.4	0.7	78.5		WATER
42	92.5	3640	31.0	0.6	71.8		WATER

COMPLETION COREGRAPH

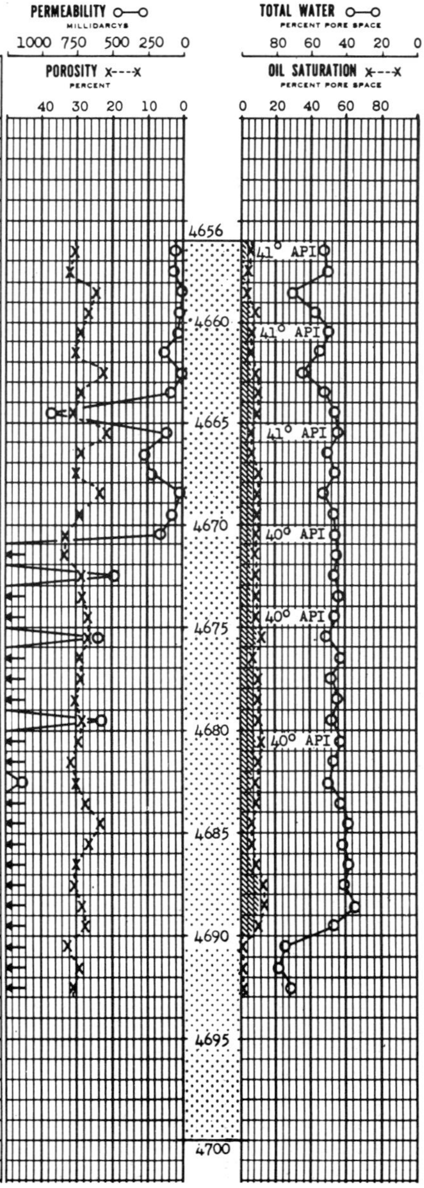

Fig. 53.533. (from *Corelab*).

SINTON SAND, TEX.

SAND · SHALE · LIMESTONE · DOLOMITE · CONGLOMERATE · CHERT

TABULAR DATA and INTERPRETATION

SAMPLE NUMBER	DEPTH FEET	PERM. MD.	POROSITY %	RESIDUAL SATURATION % PORE SPACE OIL	TOTAL WATER	Vertical Perm.	PROD
1	5523-24	0.0	19.0	0.0	66.0	0.0	
2	24-25	92	25.9	0.0	53.5	11	GAS
3	25-26	65	25.2	0.0	55.1	5	GAS
4	26-27	551	30.8	0.0	42.2	72	GAS
5	27-28	846	32.3	0.0	38.0	321	GAS
6	28-29	365	30.0	0.0	38.3	216	GAS
7	29-30	433	28.6	0.0	39.1	192	GAS
8	5530-31	238	29.7	1.0	40.4	146	GAS
9	31-32	58	25.6	4.2	50.2	2.5	GAS
10	32-33	240	30.1	22.2	48.2	118	OIL
11	33-34	0.0	22.6	13.3	54.0	0.0	
12	34-35	68	26.0	17.8	48.4	4.1	OIL
13	35-36	89	24.6	22.0	51.5	22	OIL
14	36-37	575	30.7	27.2	43.9	195	OIL
15	37-38	880	32.1	26.1	37.1	437	OIL
16	38-39	0.0	21.1	0.0	78.0	0.0	
17	39-40	4540	34.4	32.3	28.2	1850	OIL
18	5540-41	4970	33.3	29.1	29.8	3260	OIL
19	41-42	5780	38.4	33.8	24.0	2200	OIL
20	5543-44	1665	34.6	25.2	35.3	820	OIL
21	44-45	452	32.3	27.3	35.0	156	OIL
22	45-46	424	31.8	23.0	42.8	375	OIL
23	46-47	531	33.2	24.7	41.3	420	OIL
24	47-48	572	32.6	20.3	41.6	263	OIL
25	48-49	353	36.9	22.8	39.6	95	OIL
26	49-50	185	31.2	18.6	53.5	47	OIL
27	5550-51	429	35.1	22.5	41.6	185	OIL
28	51-52	715	35.3	22.7	42.2	627	OIL
29	52-53	2220	36.0	17.0	46.1	1920	OIL
30	53-54	1138	34.3	17.8	38.2	685	OIL
31	54-55	1732	33.1	20.9	45.3	1140	OIL
32	55-56	1740	35.9	22.6	39.5	1560	OIL
33	56-57	992	33.9	32.4	29.2	920	OIL
34	57-58	231	30.9	12.3	51.7	13	TRANS.
35	58-59	1131	35.0	14.0	49.4	245	TRANS.
36	59-60	270	34.8	10.6	61.7	310	TRANS.
37	5560-61	167	31.7	13.9	54.5	125	TRANS.
38	61-62	160	34.3	6.4	67.7	6.0	WATER
39	62-63	1130	37.2	11.8	59.5	251	WATER
40	63-64	67	26.4	0.0	78.9	3.6	WATER
41	64-65	917	34.3	3.8	64.2	624	WATER
42	65-66	88	33.6	5.0	66.1	19	WATER
43	66-67	123	29.5	6.1	62.2	24	WATER
44	67-68	420	32.1	0.0	68.0	28	WATER
45	68-69	12	24.0	0.0	79.2	0.8	WATER
46	69-70	0.0	23.5	0.0	82.0	0.0	WATER

COMPLETION COREGRAPH

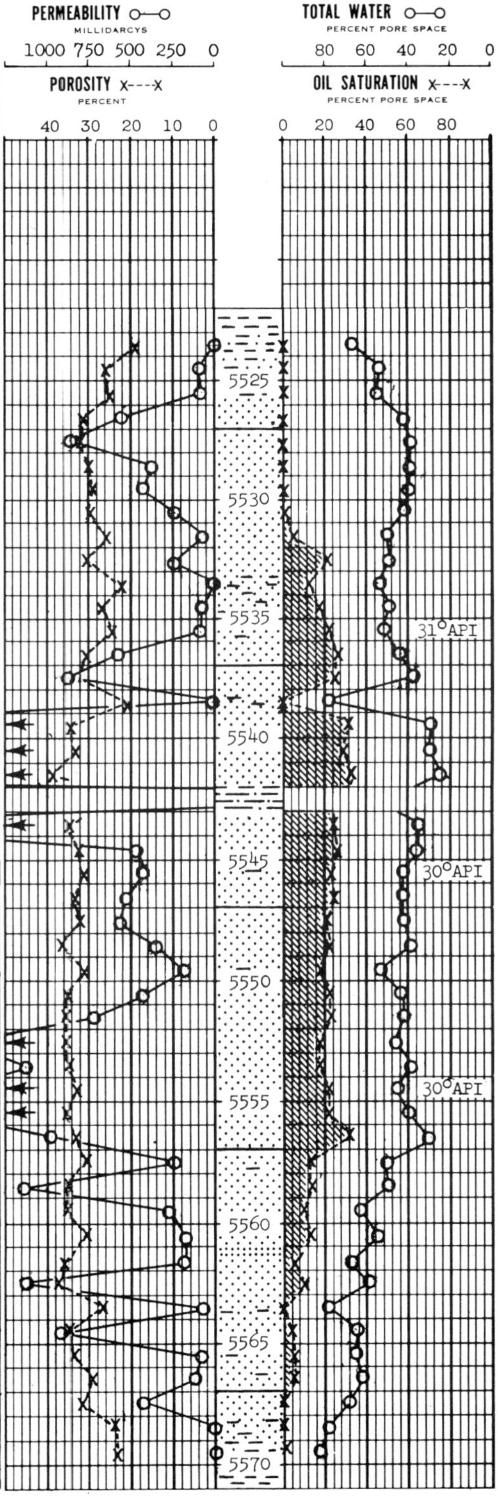

Fig. 53.534. (from *Corelab*).

The following can thus be calculated:

(a) The average permeability value in the interval (either arithmetic or geometrical mean).

(b) The average values in the interval for:

. Storage capacity: $\phi \times h$.
. Production capacity: $k \times h$.
(k, ϕ have their usual meanings: permeability, porosity and h is the productive depth).

The values obtained in this way can be compared with those obtained with other methods: tests during drilling, logging, etc.

(c) Connate water saturation either by direct measurement (oil sludge) or by calculation (empirically on the basis of total water) or by capillary pressures or by estimation.

(d) When certain results of thermodynamic analyses (PVT) such as GOR (gas/oil ratio) and FVF (formation volume factor) are known it is possible to estimate:

. The reserves in place.
. Maximum recovery.
. Production rates.
. Water injection rates.

55. APPLICATIONS OF CORE ANALYSIS

55.1. Exploration

(a) Exploration of structures and determination of their physical characteristics.

(b) Estimate of production possibilities for wildcats, extension wells and edge wells.

55.2. Well completion and workover operations

(a) Selection of intervals for testing.

(b) Interpretation of tests during drilling –Comparison of results– Explanation of test anomalies etc.

(c) Determination of the best combinations for order of completions when there are several horizons.

(d) Selection of intervals and choice of depths if plugs, packers, cement plugs etc. are installed to keep out water and gas influxes.

(e) Selection of intervals for perforations or acidizing.

(f) Estimation of completion efficiency.

(g) Selection of intervals for recompletion.

55.3. Field development

(a) Determination of optimal spacing.

(b) Determination of the location of new wells.

(c) Definition of field boundaries.

(d) Estimate of production for determination of field equipment.

(e) Definition of contact zones for the various fluids.

(f) Structural and stratigraphic correlations.

(g) Sampling and bases of interpretation for other well logging.

(h) Selection of intervals for optimum completion.

55.4. Well and reservoir evaluation

(a) Determination of net pay zone.

(b) Estimate of initial productivity.

(c) Estimate of water production rates and injection pressures.

(d) Estimate of decompression zones invaded by water or gas, and simultaneous production of various zones.

(e) Estimate of probable recovery.

(f) Estimate of oil or gas reserves in place.

(g) Estimates for equitable shares in unitization operations.

(h) Reservoir engineering and programming for maintaining pressure or secondary recovery.

(i) Forecasts for optimum well completion and maximum future recovery.

references

1 ALBA, P., "L'étude des roches-réservoirs à la Régie Autonome des Pétroles". *Rev. Inst. Franç de Pétrole,* 1958, Vol. XIII, n° 6, p. 950-984.

2 ALBERT, P. and BUTAULT, L., "Etude des caractéristiques capillaires du réservoir du Cap Bon par la méthode Purcell". *Rev. Inst. Franç. du Pétrole,* 1952, Vol. VII, n° 8, p. 250-266.

3 "API Recommended Practice for Core Analysis Procedure". *API* RP 40, August 1960.

4 BROOK, C.S. and PURCELL, W.R., "Surface Area-Measurements on Sedimentary Rock". *Petroleum Technology,* Vol. XII, 1952.

5 CLARAC, E., MONICARD, R. and RICHARD, L., "Techniques employées pour l'étude et la réalisation de stockages souterrains de gaz dans les réservoirs aquifères." 5 th World Petroleum Congress, New York, 1959 ; Section VIII, Paper 9.

6 *Cours Core Laboratories Inc.* : A Course in the Fundamentals of Core Analysis.

7 Cours de Production de *l'Ecole Nationale Supérieure du Pétrole et des Moteurs* (*ENSPM*): HOUPEURT, A., Tome I: *Etude des roches-magasins.* Ref. IFP 1 093, octobre 1956. HOUPEURT, A., Tome III: *Mouvements des fluides dans les gisements d'hydrocarbures. Essai des puits".* Ref. IFP 2073, septembre 1958.

8 FATT I. and DAVIS, D.H., "Reduction in Permeability with Overburden Pressure", *Trans. AIME,* 1952, p. 329.

9 GEERTSMA, J., "The Effect of Fluid Pressure Decline on Volumetric Change of Porous Rocks". *Trans. AIME,* Vol. 210, 1957.

10 HAAS, A. and MOLLIER, M. "Un aspect de calcul d'erreur sur les réserves en place d'un gisement: l'influence du nombre et de la disposition spatiale des puits". *Rev. Inst. Franç. de Pétrole,* 1974, Vol. XXIX, n° 4, p. 507-527.

11 HOUPEURT, A., "Sur l'écoulement des gaz dans les milieux poreux". *Rev. Inst. Franç. du Pétrole,* 1959, Vol. XIV, n° 11, p. 1 468-1 497; n° 12, p. 1 637-1 684.

12 HOUPEURT, A., "Etude analogique de l'écoulement radial circulaire transitoire des gaz dans les milieux poreux". *Rev. Inst. Franç. du Pétrole,* 1953, Vol. VIII, n° 4, p. 129-151; n° 5, p. 193-222; n° 6, p. 248-276.

13 HOUPEURT, A. and COURAUD, G., *Détermination de la teneur en eau interstitielle des roches-magasins à l'aide du désorbeur semi-perméable.* IFP report, 1951.

14 HOUPEURT, A., "Réflexions sur l'état d'équilibre des gisements vierges et sur les conditions mécaniques de la migration du pétrole". *VII Convesno Nazionale del Metane e del Petroleo,* Taormina, 1952.

15 HOUPEURT, A., ALBERT, P. and MANASTERSKI, G., *Mesure précise du volume total des échantillons de roche, en vue de la détermination de leur porosité.* (CT n° 18 du département physico-chimie des gisements, *Institut Français du Pétrole,* mars 1950).

16 HOUPEURT, A., *Mécanique des fluides dans les milieux poreux. Critiques et recherches.* Editions Technip, Paris, 1974.

17 IFFLY, R., "Etude de l'écoulement de gaz dans les milieux poreux. Application à la détermination de la morphologie des roches". Thèse doctorat. *Rev. Inst. Franç. du Pétrole,* 1956, Vol. XI n° 6, p. 757-796; n° 8, p. 975-1018.

18 JENKINS, R.E. and KOEPF, E.H., "An Evaluation of Side Wall Core Analysis Techniques". *Trans. AIME,* New Orleans, 17 Feb. 1957.

19 HUBBERT, K., "Entrapment of Petroleum under Hydrodynamic Conditions". *AAPG,* Vol. 37, n° 8, Aug. 1953.

20 KLINKENBERG, L.J., *The Permeability of Porous Media to Liquids and Gases.* Drilling and Production Practice, 1941.

21 KRUMBEIN, W.C. and SLOSS, L.L., *Stratigraphy and Sedimentation,* p. 218, Appleton Century. Crofts Inc, New York.

22 LE TIRANT, P., GAY, L., KERBOURC'H, P., MOULINIER, J. and VEILLON, D., *Manuel de fracturation hydraulique.* Editions Technip, Paris, 1972.

23 MARLE, C., Cours de production *Ecole Nationale Supérieure du Pétrole et des Moteurs (ENSPM),* Tome IV, *Les écoulements polyphasiques en milieu poreux.* Editions Technip, Paris, 1972.

24 MATHERON, G., "Structure et composition des perméabilités". *Rev. Inst. Franç. du Pétrole,* 1966, vol. XXI, n° 4, p. 564-580. "Composition des perméabilités en milieu poreux hétérogène. Méthode de Schwydler et règles de pondération". *Rev. Inst. Franç. du Pétrole,* 1967, Vol. XXII, n° 3, p. 443-446.

25 METROT, R., "Perméamètres de construction simple". *Rev. Inst. Franç. du Pétrole,* 1946, Vol. I, n° 2, p. 79-88.

26 PERRODON, A., *Géologie du pétrole.* Presses Universitaires de France, 1966.

27 *Petroleum Production Handbook* (Vol. I and II), Frick T.C. ed., McGraw Hill Book Company, 1962.

28 RAIGA-CLEMENCEAU, J., "Interprétation des diagraphies. Analyse quantitative continue de la saturation en eau. Approche rapide simplifiée". *Revue AFTP,* mars-avril 1974, n° 224, p. 17-24.

29 *Recherche sur les méthodes d'analyse des sables peu consolidés.* Ref. IFP., 942, mai 1956.

30 REUDELHUBER, F.O. and FUREN, J.E., "Interpretation and Application of Side Wall Core Analysis Data. *Transactions Gulf Coast Ass. of Geol. Societies,* Vol. VII, 1957.

31 ROCKWOOD, S.H., *Analysis of Unconsolidated Core Samples.* Drilling and Production Practice, 1948.

32 VAN DER KNAAP, W., "Non-linear Elastic Behavior of Porous Media" (presented before Society of Petroleum Engineers). *Trans. AIME,* Houston, Texas, Oct. 1958.

33 WISENBAKER, J.D., "Quick Freezing Seals Fluids Contents in the Cores". *Petrol. Engr.,* Jan. 1947, Vol. 18, p. 75-84; *Oil and Gas Journal,* 25 Jan. 1947, p. 275-281.

Subject Index

Analogies with Darcy's law, 71, 72

Archie's law, 26, 117, 118

Applications of core analysis, 160, 161

Boyle's law porosimeter, 20, 21

Capillary migration, 97, 98

Capillary pressure
- average results and correlations, 113, 114, 115
- conversion of laboratory results, 112, 113
- definition (in field), 90, 91, 92

Capillary properties of the rocks, 87, 88 89, 90

Connate water, 86, 115

Conventional core analysis, 140, 141, 142

Core analysis
- calculations, 156, 161
- classifications, 139, 140
- conventional, 140, 141, 142
- conventional by summation of fluids, 142, 143, 144, 145
- qualitative interpretation, 150, 151, 152, 153, 154
- quantitative interpretation, 151, 152
- side wall core analysis, 146, 147
- whole core analysis, 148, 149, 150

Effect of compaction on permeability, 66

Effect of compaction on porosity, 34, 35, 36, 37, 38

Effect of a pressure gradient, 101, 102

Factors affecting interfacial tension, 90

Final field equilibrium, 99, 100

Formation factor, 26, 27, 117

Gas expansion method, 20, 21, 22

Gas saturation: determination, 121, 144, 146, 148

Grain density, 18

Grain volume determination, 17

Immersion method, 17

Interstitial water by core analysis, 120, 121, 122, 123, 124, 125, 126, 145, 146, 148
- definition, 86
- determination by centrifuging, 111
  - by evaporation, 110
  - by mercury injection, 108, 109
  - by restored state method, 103, 104, 105, 106, 107
  - in the laboratory, 101
- in situ determination, 117, 118, 119

Irreducible water saturation and permeability porosity, 115, 116, 117

Klinkenberg effect, 45, 46, 47

Macroporous media, 25

Measurement media,
- pore volume, 20, 21, 22, 23, 24
- solid volume, 15, 17
- total porosity, 24
- total volume, 13, 14

Mechanical characteristics of some sedimentary rocks, 38, 40

Messer method (evaporation), 110

Morphological study of pores by mercury injection, 94, 95, 96, 97

Natural rock radioactivity measured in the laboratory, 127 to 137

Oil saturation: determination, 121 to 126, 144 to 150

Pentane cut method, 146, 147

Permeability
- average for several layers, 60, 61, 62, 63, 64
- Darcy's law, 43, 44, 50, 51
- definition, 43
- indirect determination, 81, 82, 83
- laboratory measurement, 73 to 81
- specific, effective, relative, 47, 49
- units, 44, 45

Permeability, internal surface areas, correlation, 83

Permeability, porosity, correlation, 83, 84

Porosity
- accuracy, 9, 11
- compression chamber, 18
- definition, 7, 8
- fluid summation method, 12, 142 to 147
- gas expansion, 20, 21, 22
- immersion method, 17
- indirect method (logging), 26 to 33
- by resaturation, 23
- total, 24

Porous media, 2

Preparation of samples, 12

Preservation (core analysis), 140, 141

Purcell method, 108, 109

Real density values, (*see* Grain density, 18)

Relation between permeability and other parameters, 69, 70, 71

Relation between permeability and pore morphology, 67, 68

Resaturation porosity procedure, 23

Resistivity ratio, 117, 118

Restored state method, 103 to 107

Retort method, 144 to 149

Sampling (core analysis), 141

Shape parameter, 55

Side wall core analysis, 146, 147, 148

Special core analysis, classification, 140

Threshold displacement pressure, 92, 93, 94

Tortuosity, 69, 70

Unconsolidated media, 25

Use of picnometer, 15

Vacuum saturation of samples, 15

Validity of Darcy's law, 50 to 59

Vesicular media, 25

"Voltomètre", 14

Water saturation, determination by core analysis, 120 to 123, 143 to 150

Water saturation, in situ determination, 117, 118, 119

Wettable reservoirs, 93

Whole core analysis, 148, 149, 150.

ACHEVE D'IMPRIMER
EN JUILLET 1980
PAR L'IMPRIMERIE LOUIS-JEAN
05002 – GAP
N° d'éditeur 526 – N° d'impression 408-1980
Dépôt légal : 3e trimestre 1980

IMPRIME EN FRANCE